W9-BMY-665

WETLAND
ECOSYSTEM
TEAM

UNIVERSITY OF WASHINGTON

The Estuarine Ecosystem

The Estuarine Ecosystem

Ecology, Threats, and Management

Donald S. McLusky
University of Stirling, Scotland, UK

Michael Elliott
University of Hull, England, UK

THIRD EDITION

OXFORD
UNIVERSITY PRESS

OXFORD

UNIVERSITY PRESS

Great Clarendon Street, Oxford OX2 6DP

Oxford University Press is a department of the University of Oxford.
It furthers the University's objective of excellence in research, scholarship,
and education by publishing worldwide in

Oxford New York

Auckland Bangkok Buenos Aires Cape Town Chennai
Dar es Salaam Delhi Hong Kong Istanbul Karachi Kolkata
Kuala Lumpur Madrid Melbourne Mexico City Mumbai Nairobi
São Paulo Shanghai Taipei Tokyo Toronto

Oxford is a registered trade mark of Oxford University Press
in the UK and in certain other countries

Published in the United States
by Oxford University Press Inc., New York

© D.S. McLusky 1981

© Blackie & Son Ltd. 1989

© Oxford University Press 2004

The moral rights of the author have been asserted
Database right Oxford University Press (maker)

First edition published by Blackie & Son, Glasgow 1981
Second edition published in the United States by Chapman & Hall, New York;
in Great Britain by Blackie & Son, Glasgow 1989

Third edition published by Oxford University Press 2004

All rights reserved. No part of this publication may be reproduced,
stored in a retrieval system, or transmitted, in any form or by any means,
without the prior permission in writing of Oxford University Press,
or as expressly permitted by law, or under terms agreed with the appropriate
reprographics rights organization. Enquiries concerning reproduction
outside the scope of the above should be sent to the Rights Department,
Oxford University Press, at the address above

You must not circulate this book in any other binding or cover
and you must impose this same condition on any acquirer

A catalogue record for this title is available from the British Library

Library of Congress Cataloging in Publication Data
(Data available)
ISBN 0 19 853091 9 (Hbk)
ISBN 0 19 852508 7 (Pbk)

10 9 8 7 6 5 4 3 2 1

Typeset by Newgen Imaging Systems (P) Ltd., Chennai, India
Printed in Great Britain
on acid-free paper by Antony Rowe Ltd, Chippenham, Wiltshire

Preface

For the inhabitants of many of the world's major cities and towns, estuaries provide their nearest glimpse of a natural habitat; a habitat which, despite the attempts of man to pollute it or reclaim it, has remained a fascinating insight into a natural world where energy is transformed from sunlight into plant material, and then through the steps of a food chain is converted into a rich food supply for birds and fish. The biologist has become interested in estuaries as areas in which to study the responses of animals and plants to severe environmental gradients. Gradients of salinity, for example, and the problems of living in turbid water or a muddy substrate, prevent most animal species from the adjacent sea or rivers from entering estuaries. In spite of these problems, life in estuaries can be very abundant because estuarine mud is a rich food supply, which can support a large number of animals with a large total weight and a high annual production. Indeed estuaries have been claimed to be among the most productive natural habitats in the world.

When the previous editions of this book appeared, biologists were beginning to realize that the estuarine ecosystem was an ideal habitat in which to observe the processes controlling biological productivity. In the intervening period, several more estuaries and their inhabitants have been studied intensively, and it is now possible to answer many of the questions posed by the earlier edition, and to pursue further the explanation of high productivity in estuaries and of energy utilization at different trophic levels within the estuarine food web.

Users of the previous editions were kind enough to welcome the framework of the book, which first outlined the estuarine environment and the physical and biological factors, which are important within it. We then examine the responses of the animals and plants to these factors, consider the problems of life in estuaries and why so few species have adapted to estuaries, and then propose a food web for an estuary. Thereafter we shall examine each trophic level in the food web in turn, first the primary producers (plants and detritus), then the primary consumers (herbivores and detritivores), and finally the secondary consumers (carnivores). These chapters have been fully revised in this third edition, to reflect our latest knowledge in these topics.

In the period since the publication of the previous editions of this book, a vast amount of new information on pollution in estuaries has accumulated. It has been widely recognized that although the world's seas are huge and may appear capable of receiving unlimited quantities of man's waste, such waste is often discharged first into the confined waters of estuaries. Many international experts have stated that, while the open oceans may not be generally polluted, the coastal waters of the sea and especially the waters of estuaries are widely polluted. Thus in practice, marine pollution is often essentially estuarine pollution. To reflect this large impact of mankind on estuaries, and to consider how mankind may either destroy or enrich the estuarine ecosystem, new chapters have been prepared in this edition. These consider pollution in estuaries, and the diverse uses and abuses of the estuarine habitat by man, as well as the methods used to study human induced changes in estuaries and the ways in which estuarine management can either monitor, control or prevent pollution or destruction of the estuarine ecosystem.

This latest edition therefore retains the concept of the study of the ecosystem as the basis for our understanding of the natural world, and shows that estuaries are ideal habitats for such studies. The new content and chapters reflect our attempts to recognize both the problems of pollution in estuaries and the solutions, which estuarine management can offer, as estuarine ecosystems come under increasing pressure from a wide range of demands made by an increasing world population. The topic of estuarine pollution, estuarine uses and abuses, and their management has moved on considerably in the past decade. As many of the problems of "point" source pollution have been attended to, different problems of "diffuse" pollution have become apparent. In addition, while some pollution problems are seen as capable of being solved, habitat loss and degradation is more of a threat. Hence the third edition concentrates on these as well as including sections on the effects of fisheries, dredging, structures, aquaculture, etc. on estuarine ecosystems. The third edition seeks to include considerably more applied material and present case studies of estuarine change.

Estuaries can be perceived as either the originators of pollutants, or more often as the recipients of pollution that originate on land, freshwater, or from the sea. The problems of estuarine pollution and management are those dealing with a sheltered environment that acts as a trap for sediment and contaminants, and as areas subject to intense pressure from mankind's activities such as land-claim (or reclamation).

Estuaries are the most protected habitat in the United Kingdom and elsewhere (measured as the percentage of the total habitat that is subject to protection orders), and for most students on aquatic biology are their most accessible marine habitat. While few UK, European, or American Universities have ships for access to the sea, many are located on the shores of estuaries. Thus for many University courses, their local estuary is the marine habitat, which can be studied, while the open seas have to be studied only in films or video. Throughout the book examples are drawn from Britain, Europe, and America, as well as other areas, which have been well studied such as South Africa and Australia.

May we thank the following authors and publishers for allowing us to use copies of their figures: Kai Olsen (Ophelia; 3.3, 4.1), Colin Moore (Royal Society of Edinburgh; 4.8), Sven Ankar and Ragnar Elmgren (Askö Contributions; 4.10), Peter Burkhill (Plymouth Marine Laboratory; 4.13), Tom Pearson, and Rutger Rosenberg (Oceanography and Marine Biology, Annual Review; 6.6, 6.7), Steve Colclough (Environment Agency, London; 6.9, 6.26), John Beardmore (Swansea University; 7.5), Blackwells Scientific Publishers (6.10, 6.11, 6.23, 6.24, 7.4). Jan Elliott and Krystal Hemingway (IECS, University of Hull) are thanked for preparing the complex flowcharts (the 'horrendograms'!) in Chapters 6 and 7 and Sue Boyes (IECS, University of Hull) is thanked for providing information for Chapter 8. The authors prepared other figures. Sources of information that are not mentioned in the text are listed in the reading list, which has been fully updated for this edition.

Donald S. McLusky
Michael Elliott
2004

Contents

CHAPTER 1

The estuarine environment

1.1 Introduction

Estuaries have for long been important to mankind, either as places of navigation, or as locations on their banks for towns and cities. Nowadays they are under pressure, either as repositories for the effluent of industrial processes and domestic waste, or as prime sites for land-claim to create sites for industry or urban development. Against this background the biologist has been attracted to other functions of estuaries: Vital feeding areas for many species of birds, especially waders and wildfowl, like the locations of coastal fisheries or as fascinating areas present challenges to our understanding of how animals and plants adapt to their environment. The estuarine environment is characterized by having a constantly changing mixture of salt and freshwater, and by being dominated by fine sedimentary material carried into the estuary from the sea and from rivers, which accumulates in the estuary to form mudflats. The mixtures of salt and freshwater present challenges to the physiology of the animals, which few are able to adapt to.

In this book we shall examine estuaries by trying to understand the interactions and feeding relationships, which make up the estuarine food web. Estuaries have been claimed to be amongst the most productive natural habitats in the world, and we shall attempt to explain why they are so productive, and how the energy produced is utilized by succeeding trophic levels. Estuaries are transition zones between rivers and the sea, which differ from both in abiotic and biotic conditions. Temperature, salinity, and turbidity fluctuate on a daily basis and reach more extremes in estuarine waters than they do at sea or in rivers. From a biotic viewpoint, estuaries are highly productive ecosystems ranking at the same level as coral reefs and mangrove swamps. An elevated productivity is maintained because of high nutrient levels in both sediment and water column.

Before we can examine the life of an estuary we must first examine the physical and chemical features, which mould the estuarine environment. An estuary is a partially enclosed body of water formed where freshwater from rivers and streams flows into the oceans, mixing with the seawater. Estuaries and the lands surrounding them are places of transition from land to sea, and from fresh to saltwater. Although influenced by tides, estuaries are protected from the full force of the ocean waves, winds and storms by reefs, barrier islands or fingers of land, mud, and sand that define an estuary's coastal boundary. Estuaries come in all shapes and sizes and can be called by many different names—bays, lagoons, harbors, and inlets or sounds.

When considering any estuarine habitat worldwide there are many generalizations that can be perceived, for example, common features being the gradient of conditions from the open sea into the sheltered estuary, and on to the freshwater river. Along this gradient there are clear changes in salinity ranging from full strength seawater decreasing to freshwater. Associated changes in sedimentary conditions from coarse sediment (sand or gravel) without the estuaries to fine sediments (muds) within the estuaries are invariably found. Other changes relate to alterations in turbidity of the water column, or in chemical composition including changes in nutrients, dissolved gases, and trace metals.

The English word "estuary" is of sixteenth century origin, derived from the Latin word *aestuarium* meaning marsh or channel, which is itself derived from *aestus*, meaning tide or billowing movement, related to the word *aestas* meaning summer. A widely used definition of an estuary has been given by Pritchard (1967) as: "an estuary is a semi-enclosed coastal body of water, which has a free connection with the open sea, and within which sea water is measurably diluted with fresh water derived from land drainage." This definition of estuaries excludes coastal lagoons or brackish seas, although we will discuss them briefly later in this book as useful comparisons. Coastal lagoons, for example, do not usually have a free connection with the open sea and may only be inundated with seawater at irregular intervals. Brackish seas, such as the Caspian Sea, may have salinities comparable to some parts of estuaries, but they do not experience the regular fluctuations of salinity due to the tide. The definition of "*semi-enclosed*" serves to exclude coastal marine bays and the definition of "*freshwater derived from land drainage*" serves to exclude saline lakes with freshwater from rainfall only.

While many estuarine scientists have used Pritchard's definition, studies in the tidal freshwater regions of estuaries have suggested that the definition of Fairbridge (1980) is more suitable, namely that

an estuary is an inlet of the sea reaching into a river valley as far as the upper limit of tidal rise, usually being divisible into three sectors: a) a marine or lower estuary, in free connections with the open sea; b) a middle estuary subject to strong salt and freshwater mixing; and c) an upper or fluvial estuary, characterized by freshwater but subject to strong tidal action. The limits between these sectors are variable and subject to constant changes in the river discharges

The principal difference between the two definitions is in determining the upper limit.

For Pritchard it is the upstream limit of salt penetration, and for Fairbridge it is the upstream limit of tidal penetration, which in an un-modified estuary will always be further inland. The Fairbridge definition also usefully emphasizes the gradient of conditions that may be found in a normal estuary.

The salinity of seawater is approximately 35, tending to be lower (33 in coastal seas and higher 37) in tropical waters. The salinity of freshwater is always less than 0.5. Thus the salinity of estuarine waters is between 0.5 and 35. This range is generally termed brackish, as distinct from marine or freshwaters. Salinity is a measure of the salt content of the water, and is expressed throughout this book using the Practical Salinity Scale (see Box 1.1). Whereas marine and freshwaters are characterized by stable salinities, estuarine water is extremely variable in its salinity.

An estuary is thus emphasized as a dynamic ecosystem having a free connection with the open sea through which seawater normally enters according to the twice-daily rhythm of the tides. The seawater that enters the estuary is then measurably diluted with freshwater flowing into the estuary from rivers. The patterns of dilution of the seawater by freshwater vary from estuary to estuary depending on the volume of freshwater, the range of tidal amplitude, and the extent of evaporation from the water within the estuary and may be used as a basis for the classification of estuaries.

Three main types of estuaries can be recognized, namely positive, negative, or neutral estuaries. In positive estuaries the evaporation from the surface of the estuary is less than the volume of freshwater entering the estuary from rivers and land drainage. In such a positive estuary (Fig. 1.1) the outgoing freshwater floats on top of the saline water, which has entered the estuary from the sea, and water gradually mixes vertically from the bottom to the top. This type of estuary, which is the most typical in the temperate parts of the world, is thus characterized by incoming saltwater on the bottom, with gradual vertical mixing

Box 1.1 Salinity

Throughout this book salinity is expressed on the practical salinity scale, using what are called Practical Salinity Units (PSU). Salinity is a scale and is thus a dimensionless unit (like pH). In earlier editions of this book, and many other older publications, salinity was expressed as parts per thousand (often abbreviated to ppt, ‰, or $g\,kg^{-1}$) being the total concentration of salts in grams contained in 1 kg of seawater. The salts are principally sodium and chloride ions, supplemented by potassium, calcium, magnesium, and sulfate ions, plus minute or trace amounts of many other ions.

For the past two decades salinity has been defined solely in terms of electrical conductivity. Conductivity units are unfamiliar to some marine scientists and therefore have been converted to a practical salinity scale. The salinity scale without units has now been adopted as the practical scale since the numerical values are closely similar to the values previously expressed in parts per thousand of salt, the only obvious difference being the lack of any symbol. Since salinity is now expressed as a scale, it is correct to say "The salinity of seawater is 34," but it is incorrect to say "34 PSU, ppt, ‰, or $g\,kg^{-1}$."

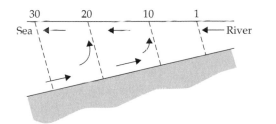

Figure 1.1 Positive estuary. Freshwater runoff is greater than evaporation. The arrows show the pattern of circulation with denser marine water entering the estuary along the bottom and then gradually mixing vertically with the outgoing surface stream of fresher water. Salinity is expressed in PSU (see Box 1.1).

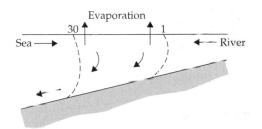
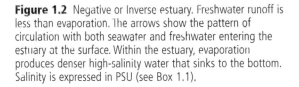

Figure 1.2 Negative or Inverse estuary. Freshwater runoff is less than evaporation. The arrows show the pattern of circulation with both seawater and freshwater entering the estuary at the surface. Within the estuary, evaporation produces denser high-salinity water that sinks to the bottom. Salinity is expressed in PSU (see Box 1.1).

leading to an outgoing stream of fresher surface water.

In negative (or inverse) estuaries the opposite situation exists, with evaporation from the surface exceeding the freshwater runoff entering the estuary. This type of estuary is mostly found in the tropics, for example, the Laguna Madre, Texas, or Spencer Gulf and Gulf St Vincent, South Australia, although it can also occur in temperate regions where the freshwater input is very limited such as the Isefjord in Denmark. The Mediterranean and Adriatic Seas are regarded as prime examples of inverse estuarine systems, as well as the Arabian Gulf, the Red Sea. In negative estuaries evaporation

causes the surface salinity to increase. This saltier surface water is then denser than the water underneath and thus sinks. The circulation pattern is thus opposite to that of a positive estuary, because in a negative estuary the sea and freshwater both enter the estuary on the surface, but after evaporation and sinking they leave the estuary as an outgoing bottom current (Fig. 1.2).

Occasionally the freshwater input to the estuary exactly equals the evaporation and in such situations a static salinity regime occurs (Fig. 1.3). Such an estuary is termed a neutral estuary, but they are rare, as evaporation and freshwater inflow are almost never equal.

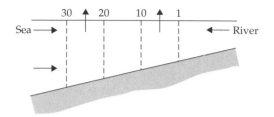

Figure 1.3 Neutral estuary. Freshwater runoff is equal to evaporation. This situation is intermediate, and possibly transitory, between a positive estuary (Fig. 1.1) and a negative estuary (Fig. 1.2). Salinity is expressed in PSU (see Box 1.1).

Table 1.1 The proportions of the British estuarine resources areas within estuaries of different geomorphologic types—see Box 1.2

Estuary type	% of total area	% of intertidal area
Fjord	2	1
Fjard	5	6
Ria	3	2
Coastal plain	35	31
Bar-built	6	8
Complex	18	17
Barrier beach	2	3
Linear shore	4	6
Embayment	25	26

Source: Davidson *et al.* (1991).

In this book the emphasis throughout will be on the commoner positive type of estuary. In general, estuaries are a habitat intermediate between the sea, the land, and freshwaters: A habitat, which is a complex dynamic mixture of transitional situations and is almost never static and where physical and chemical factors show marked variations. Many of these factors are closely linked to the salinity distribution patterns mentioned above, such as the strength of currents, tidal amplitude, wave strength, and deposition of sediments as well as temperature, oxygen, and the supply of nutrients.

Just as the physical and chemical factors within estuaries are in a continual state of change, so even the topography of estuaries is continually changing. Most estuaries have developed on the low-lying ground that forms the coastal plains around landmasses. Here estuaries typically develop in river valleys that become drowned by the sea. In the drowned valleys extensive deposition of sediments carried into the estuaries have produced the estuaries that we see today. The various types of estuaries are described in Box 1.2, with their frequency in the British Isles, as an example, seen in Table 1.1. Many estuaries are drowned river valleys that have become estuaries since the last Ice Age, which is within the last 10–15,000 years. The height of the sea in relation to the land is continually changing, and while the mean sea level has

increased as ice caps have melted, so in some areas the land has also subsided and in other areas the land that was formerly glaciated has sprung up as the weight of ice on it has melted.

Many estuaries of South Africa and Australia are of the bar-built type, but in these countries the sandbars may seal the estuary from the sea during the drier months of the year. The estuary is thus only opened when the sand bar is breached either by high riverine discharge or by artificial means (such as a dredger). During the often lengthy periods when they are closed these so-called "blind" estuaries do *not* have a free connection with the open sea, considered to be crucial for the Pritchard and Fairbridge definitions of an estuary. In order to accommodate the peculiarities of these systems Day (1980) redefined an estuary as "a partially enclosed coastal body of water which is either permanently or periodically open to the sea and within which there is a measurable variation of salinity due to the mixture of sea water with freshwater derived from land drainage."

As siltation of estuaries occurs and salt marshes extend and consolidate the land, so the waters of the estuary may be pushed seawards. Small changes in sea level due to long-term climatic change could either drown or expose many of our present estuaries,

---

Box 1.2 Types of estuaries (after Fairbridge and Davidson *et al.* 1991)

Fjords. Drowned glacial troughs. Fjord-type estuaries occur where valleys have been deeply eroded by glaciation. Characterized by deep inner basins linked to the sea by shallow entrance sills, for example, Sea lochs in West of Scotland, Fjords in Norway, Sweden, Alaska, British Columbia, New Zealand, such as the coasts of Norway, Western Scotland, Alaska, and New Zealand.

Fjards. Typical of glaciated lowland coasts. More complex than fjords, with a more open and irregular coastline, for example, Solway Firth England/Scotland, eastern Canada, and New England.

Rias. Drowned river valleys, formed by subsidence of land and/or a rise in sea level. Deep, narrow channels with a strong marine influence, for example, Estuaries of Cornwall, England and Brittany, France.

Coastal plain estuaries. Formed by the flooding of pre-existing valleys. Unlike Rias, these estuaries are often very shallow and filled with sediment so that extensive mudflats and saltmarshes occur. Commonest type of estuary in United Kingdom, for example, Severn, Dee, Humber, Thames, England. Chesapeake Bay, Charleston Harbor, Delaware Bay, USA.

Bar-built estuaries. Also drowned river valleys, but recent sedimentation has kept pace with their drowning so that they have a characteristic bar across their mouths, for example, Alde, England; Ythan, Scotland. Barnegat Bay, New Jersey, Laguna Madre, Texas, Albufeira, Portugal, and most estuaries of North Carolina—Florida coast. In many estuaries in South Africa and Australia the bar may seasonally close the estuary, creating closed or blind estuaries.

Complex estuaries. Drowned river valleys of complex origin, typically a mixture of glaciation, river erosion, and sea level rise, for example, Scottish Firths: Solway, Moray, Dornoch, Tay, and Forth. San Francisco Bay is a complex estuary created by tectonic activity (land movement due to faulting).

Barrier beaches. Open coast system where a bar or barrier develops offshore, and an estuary is thereby created behind the barrier. For example, North Norfolk Coast, Lindisfarne, England.

Linear shore sites. Formed where the shore is sheltered, for example, by barrier islands. Usually considered as a subdivision of a complex estuary. For example Essex and North Kent coast, England.

Embayments. Large natural areas formed between rocky headlands that naturally fill with soft sediments. For example, Carmathen bay, Wales, Morecambe bay, The wash, England.

---

although of course new ones may be formed elsewhere. The various types of estuary, however formed, are all characterized by the variability of their environmental parameters, most notably in their water circulation and their sediments.

1.2 Estuarine circulation

Within the commonest "positive" estuaries four main types can be recognized depending on the tidal amplitude and volume of fresh-water flow (Dyer 1997):

1. *Salt wedge or Highly stratified*. Examples—Mississippi (US), N. Esk (UK), Vellar (India)

2. *Fjord*. Examples—Norwegian fjords, West of Scotland sea lochs. Silver Bay (Alaska), Alberni Inlet (British Columbia)

3. *Partially mixed*. Examples—James River (US), Mersey, Thames, Humber, Forth (UK), Elbe (Germany)

4. *Homogeneous*. Examples—Delaware, Raritan (US), Solway Firth (UK), Mandovi-Zuari (India)

In the first of these, the highly stratified or salt wedge, the freshwater flows seawards on the surface of the inflowing saltwater. At the inter-face between the salt and freshwater, entrain-ment (mixing) occurs and saltwater is mixed into the outflowing freshwater (Fig. 1.4). For

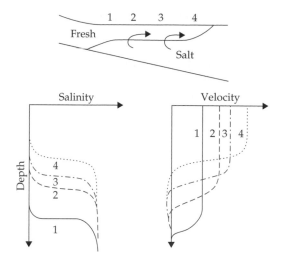

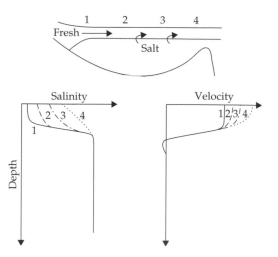

Figure 1.4 Salinity and velocity profiles in a salt wedge estuary. The salinity and velocity profiles for four positions in the estuary are shown, with arbitrary axes. (After Dyer 1997.)

Figure 1.5 Salinity and velocity profiles in a Fjord estuary. The salinity and velocity profiles for four positions in the estuary are shown, with arbitrary axes. (After Dyer 1997.)

this type of estuary a large river flow in relation to tidal flow is needed and continuous down-stream flow of surface water may occur despite the ebb and flood of the salty tidal water beneath it.

The second type, the fjord, is basically simi-lar to the highly stratified, except that due to a sill (a shallow lip) at the mouth of fjords the inflow of tidal water is more restricted. Again a continuous downstream flow of freshwater at the surface occurs, but the renewal of tidal water may only occur seasonally and nonre-newal of water may lead to anoxic conditions in the deepest parts of the fjord (Fig. 1.5).

Given a situation with the tidal inflow greater than, or similar to, the freshwater inflow a partially mixed estuary develops. In such an estuary there is a continuous mixing between the sea and freshwater. Surface waters will be less saline than the bottom waters at any given point in the estuary, but unlike the highly strat-ified estuary undiluted freshwater will only be found near the head of the estuary. Mixing of water from the predominantly inflowing bottom to the mainly outflowing surface will occur throughout the estuary (Fig. 1.6). The

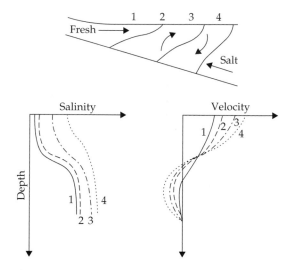

Figure 1.6 Salinity and velocity profiles in a partially mixed estuary. The salinity and velocity profiles for four positions in the estuary are shown, with arbitrary axes. (After Dyer 1997.)

pattern of mixing may be less clear at the mar-gins of the estuary, and due to the Coriolis force the seawater will dominate on the left-hand side (looking downstream) in the northern hemisphere, and the outgoing fresher water will dominate the right-hand side. In the Southern Hemisphere, this pattern is reversed.

When the estuary is very wide the Coriolis force will cause a horizontal separation of the flow, with outgoing flow on the right-hand side in the Northern Hemisphere, and ingoing flow on the left-hand side. Thus the circulation in such a homogeneous estuary will be across the estuary from left to right, rather than vertically as in the other types (Fig. 1.7).

The range of tidal amplitude varies in a constant pattern in the seas of the world and the tides within estuaries are due to the tidal wave at their mouth. Tidal amplitude varies not only from place to place but also varies at any particular locality according to the neap and spring tide cycle, as the range of tides fluctuates from a maximum rise and fall at spring tides to a lesser rise at neap tides. The full cycle from spring to neap tides occurs each lunar month (28 days). In the waters around Denmark and Sweden, for example, the tidal amplitude is very small and the many estuaries of Denmark are not subject to strong tidal currents and do not develop large intertidal areas. At the other extreme the estuaries of the Canadian Bay of Fundy region experience maximal tidal amplitude producing strong tidal currents as well as large intertidal areas.

Depending on their tidal range estuaries can be classified as microtidal, mesotidal, macrotidal, or hypertidal (Box 1.3). The tidal amplitude may considerably influence the mudflats and vegetation of an estuary. While microtidal estuaries have only limited intertidal areas, mesotidal estuaries often develop extensive intertidal areas, which are covered in vegetation, such as

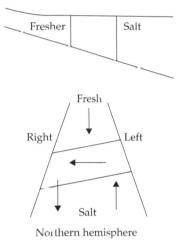

Figure 1.7 Salinity distribution in a large homogeneous estuary in the northern hemisphere. Due to the Coriolis force, horizontal separation occurs, and the circulation is across the estuary rather than vertically. (Modified after Dyer 1997.)

Box 1.3 Types of estuaries, on the basis of tidal range		
Name	Tidal range (m)	Examples
Microtidal	<2	Limfjord, Isefjord (Denmark); Breydon Water, Oulton Broad, Christchurch Harbour, Poole Harbour, The Fleet and Portland Harbour (UK); Cancun Bay (Mexico); Tampa Bay, Apalachicola Bay, Mobile bay, Mississippi, Laguna Madre, (US); Coastal lagoons
Mesotidal	2–4	Clyde, Dornoch, Cromarty, Ythan, Tay (Scotland); Orwell, Stour, Southampton, Lymington (England)
Macrotidal	4–6	Neath, Loughor, Milford Haven, Conwy (Wales); Mersey, Ribble, Morecambe bay, Tyne, Tees, Thames, Exe, Dart (England); Lossie, Forth, (Scotland); Yellow Sea (China); Delaware (US)
Hypertidal	>6	Bay of Fundy (Canada); Severn (UK); Seine, Somme (France)

Spartina, whereas macrotidal and hypertidal estuaries more typically have bare mudflats without large plants, but with microscopic microphytobenthos.

In all estuaries the strength of tidal and river currents is dissipated as the main inflow currents collide and mix with each other, and compared to both the sea and rivers, estuaries are quiet places. Due to the shelter of the land surrounding the estuary, wind-driven waves are much smaller than in the sea. This absence of large waves, coupled with the mixing of currents is vital for the deposition of estuarine sediments.

The river flow, tidal range, and sediment distributions in estuaries are continually changing and consequently estuaries are never really "steady-state" systems. With increased river flow, or during neap tides, the extent of tidal intrusion of seawater diminishes, while with spring tides or decreased river flow the tidal intrusion increases. The concept of "flushing time" is one attempt to relate the freshwater flow to the tidal range, and the volume of the estuary. The flushing time (T) is the time required to replace the existing freshwater in the estuary by the river discharge, and is calculated by dividing the total volume of river water accumulated in the estuary (Q) by the river flow (R) (Dyer 1997) thus: $T = Q/R$.

In the Forth estuary for example, the flushing time under mean river flow conditions is 12 days, but in summer with reduced river flow it may be up to 10 weeks, and following severe rainfall it may be down to 6 days. These values need, of course, to be calculated individually for each estuary. Their calculation can be of great value in predicting the impact of polluting discharges on estuaries, as an estuary with a short flushing time is generally better mixed and better able to accept effluents.

Within particular estuaries the pattern of circulation may also vary throughout the year in response to meteorological events. The Potomac estuary (US) is a partially mixed estuary for 43% of the year, with surface outflow and deep inflow, but the reverse with surface

Table 1.2 Classification of estuarine divisions

Division	Tidal	Salinity	Venice system (1959)
River Head	Non-tidal The highest point to which tides reach	<0.5	Limnetic
Tidal fresh	Tidal	<0.5	Limnetic
Upper	Tidal	0.5–5	Oligohaline
Inner	Tidal	5–18	Mesohaline
Middle	Tidal	18–25	Polyhaline
Lower	Tidal	25–30	Polyhaline
Mouth	Tidal	>30	Euhaline

Source: McLusky (1993).

inflow and deep outflow also occurs for 20% of the year, and four other patterns occur for the rest of the time in response to climatic events.

The salinity at any particular point of an estuary depends on the relationship between the volume of tidal seawater and the volume of freshwater entering the estuary, as well as the tidal amplitude, the topography of the estuary, and the climate of the locality, but in general it is possible to recognize various zones or divisions within an estuary (Table 1.2)

While the salinity of the water bodies of estuaries is important for fish and planktonic organisms living in the water column, it is of less direct importance for the majority of estuarine animals, which live buried within the muddy deposits. Far more important for these benthic animals is the interstitial salinity, which is the salinity of the water between the mud particles. It has been consistently shown that the interstitial salinity varies much less than the salinity of the overlying water, due to the slow rate of interchange between them. On the intertidal mudflats, where the most abundant populations of estuarine animals are often found, the interstitial salinity matches that of the high-tide salinity when it covers the mudflats, which may be considerably higher than the salinity of the estuarine water at low tide at the same location. Because of this phenomenon it is usually possible for marine animals living buried in the sediment to penetrate further into estuaries, than for marine animals living planktonically.

1.3 Estuarine sediments

Fine sedimentary deposits, or muds, are a most characteristic feature of estuaries, and indeed the estuarine ecosystem has been defined as "a mixing region between sea and inland water of such shape and depth that the net resident time of suspended (sedimentary) materials exceeds the flushing" (Hedgpeth 1967). Sedimentary material is transported into the estuary from rivers or the sea, or is washed in from the land surrounding the estuary. In most North European and North American estuaries the main source of sedimentary material is the sea, and the material is carried into the estuary either as suspended sediment flux or as bedload transported in the bottom inflowing currents that characterize salt wedges. In the Tay estuary, Scotland, UK, for example, 70% of the sediments accumulating on the tidal flats are of marine origin. However, in some French estuaries, where fine material is available on the banks of the

estuary, these banks are the main source of estuarine sediments. In the estuaries of the Loire (France), Vigo (Spain), Apalachicola (US) and Yellow River (China), rivers carrying large quantities of clay are the main source of estuarine sediments.

Whatever the source of the sediments the deposition of it within the estuary is controlled by the speed of the currents and the particle size of the sediments. The relationship between current speed and the erosion, transportation, and deposition of sediments is shown in Fig. 1.8. It can be seen that for a sediment with pebbles 10^4 μm (1 cm) diameter erosion of the sediment will take place at current speeds of over 150 cm s^{-1}. At current speeds between 150 and 90 cm s^{-1} the pebbles will be transported by the current, but will be deposited at current speeds less than 90 cm s^{-1}. Similarly for a fine sand of 10^2 μm (0.1 mm) diameter erosion will occur at speeds greater than 30 cm s^{-1}, and deposition will occur at speeds less than 15 cm s^{-1}. For silts and clays with particles of 1–10 μm diameter a similar relationship occurs, except that consolidation of the sediment means that for erosion to occur faster current speeds are needed as the consolidated sediment behaves as if it were composed of larger size particles.

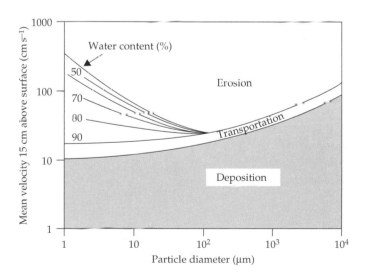

Figure 1.8 Erosion, transportation, and deposition velocities for different grain particle sizes. The diagram also illustrates the effect of the water content of the sediment on the degree of consolidation, which in turn modifies the erosion velocities. (After Postma 1967.)

The consequence of these relationships for estuaries is that in the fast-flowing rivers and strong tidal currents at either end of the estuary all sizes of sedimentary particles may be eroded and transported. As the currents start to slacken within the estuary so the coarser pebbles and sands will be the first to be deposited, and the finer silts and clays will remain in suspension. In the calmer middle and upper reaches of an estuary, where the river and tidal currents meet, and especially in the slack water at high tide overlying the intertidal areas, only then will the currents be slow enough for mud to be deposited.

The rate of deposition, or the settling velocity, of sediments is related to the particle diameter. For coarser particles (medium sand and larger) the settling velocity is determined by the size of the particle and varies as the square root of particle diameter. This relationship, known as the impact law, is represented by the equation:

$$V = 33 \sqrt{d}$$

where V is the settling velocity (cm s^{-1}) and d is the diameter of a spherical grain of quartz (cm). For finer particles (silts, clays, and fine sands) the viscous resistance of the fluid in which the particle is settling determines the settling velocity and Stokes' law applies. Particle shape, concentration, density, and efficiency of dispersion all influence the precise values for Stokes' law. Stokes' law may be represented for quartz particles in water at 16 °C as:

$$V = 8100d^2$$
(V, d as above).

The results of the impact and Stokes' law as settling velocity of various sizes of sediments are presented in Table 1.3, from which it can be seen that sands and coarser materials settle rapidly in water, and any sediment coarser than 15 μm will settle within one tidal cycle. However, for finer sediments, settling velocities are much slower and clay and silt particles less than 4 μm diameter will certainly be unable

to fall and settle within one tidal cycle. Consequently the waters of estuaries tend to be very turbid as the silt and clay particles in suspension are carried about the estuary until they eventually settle to form the mudflats that are so characteristic of estuaries. The speed of settlement can be slightly faster than portrayed in Table 1.3 as "salt flocculation" can occur. This process is caused by the clay particles in saltwater tending to adhere to each other, and as they form larger particles so they will tend to fall faster, for example, particles of 1.5 μm in the deflocculated state form flocculated particles, which behave as if they were 7 μm in diameter. Sea salts have two roles to play in the aggregation of fine particles. First, even a small amount of salt is sufficient to make suspended particles cohesive, and second, the collisions between particles prior to aggregation occur most frequently in the density gradients produced by the mixing of salt and freshwaters in estuaries. The properties of sediments may also be greatly altered by its inhabitants. In particular the sediments may become "cohesive," which especially occurs when microscopic algae (microphytobenthos) inhabit the surface of the mud. The biological production of these important microscopic plants is discussed later, but here we should note the important role

Table 1.3 Settling velocities

Material	Median diameter (μm)	Settling velocity (m day^{-1})
Fine sand	250–125	1040
Very fine sand	125–62	301
Silt	31.2	75.2
Silt	15.6	18.8
Silt	7.8	4.7
Silt	3.9	1.2
Clay	1.95	0.3
Clay	0.98	0.074
Clay	0.49	0.018
Clay	0.25	0.004
Clay	0.12	0.001

Source: King (1975).

that they play in binding the fine particulate sediments of estuaries together.

The middle and upper reaches of estuaries are thus characterized by very turbid water with poor light penetration. Within many estuaries the suspended matter in the estuaries develop so-called "turbidity maxima." The presence and magnitude of turbidity maxima are controlled by a number of factors, including the amount of suspended matter in the river or seawater, the estuarine circulation, and the settling velocity of the available material. As Fig. 1.9 shows for the Ems estuary the turbidity maxima is located at the meeting point of the river and tidal currents. The sedimentary material in the central and upper reaches of the estuary is continually added to by river discharge and tide inflow, and remains largely in suspension, before finally settling. The position of the turbidity maximum is generally determined by the contribution of suspended sediment from the seaward

end of the estuary. The inward flow of marine water and its sediment moves along the bottom of the estuary until it reaches a stagnation point where inward flow ceases. At this stagnation point the seawater and its sediment rise to mix with the surface fresher water. The position of this stagnation point, and the consequent turbidity maximum, also varies with the strength of river flow, moving upstream with low river flow, and downstream with high river flow.

In many estuaries the maximum concentration of suspended sediment occurs at low tide, as the ebbing tide washes sediment off the intertidal areas and allows the sediment in suspension to remain in the low water channel. As the tide rises, the concentration of suspended load is reduced as the flooding tide increases the volume of water in the estuary, and the sediments are carried over the intertidal areas, where settlement may occur, at high tide. In northern latitudes, ice break-up

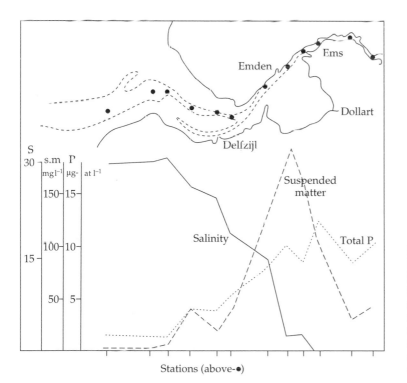

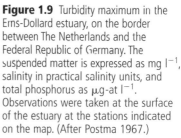

Figure 1.9 Turbidity maximum in the Ems-Dollard estuary, on the border between The Netherlands and the Federal Republic of Germany. The suspended matter is expressed as mg l^{-1}, salinity in practical salinity units, and total phosphorus as μg-at l^{-1}. Observations were taken at the surface of the estuary at the stations indicated on the map. (After Postma 1967.)

in the spring is the most significant factor affecting muddy intertidal sedimentation. During summer, organic processes dominate as organisms feed on, pelletize, and bioturbate the sediment, while plants may stabilize the sediment. In autumn (fall) seasonal storms increase, and from then into winter most sediment erosion and transportation will occur.

Along with the sediments being carried into estuaries are usually carried particles of organic debris derived from the excretion, death, and decay of plants and animals. Once the dissolved and particulate organic matter reaches estuaries from fresh and saltwater it tends to remain there as it is deposited and incorporated into the estuarine ecosystem along with fine inorganic matter. The organic matter in all aquatic ecosystems is conventionally divided into two fractions by filtration through filters with an average pore diameter of 0.5 μm. The fraction passing through the filter is termed dissolved organic matter (DOM) even though it will contain very fine particulate matter in addition to truly dissolved matter. The fraction retained by the filter is termed particulate organic matter (POM).

The sedimentation of both inorganic and organic suspended material leads to the development of mudflats and other areas of deposition within estuaries. Attempts have been made to summate the sources of suspended material, and calculate the rate of deposition of sediments. For example, within the upper Chesapeake Bay regions (US), the river supplies 83% of the suspended matter (3/4 inorganic, 1/4 organic). In the middle reaches of the same estuary, shore erosion provides 52% of the suspended matter, with production of organic matter within the estuary providing another 22%. Within the whole Chesapeake Bay system, including the lower reaches, deposition of sediment occurs at a rate of 0.8 mm year^{-1}, with 56% of the sediment supplied by rivers, 31% by shore erosion, and 12% from tidal inflow. Elsewhere in the United States, in the Patuxent River estuary, sediment collects at a rate of 3.7 cm year^{-1}, and in the large estuarine system of Long Island Sound, near New York, it has been shown that since the estuary was formed 8000 years ago the volume of marine mud supplied to the estuary greatly exceeds the volume supplied by the rivers feeding the area.

The changes in the particle size composition of sediments in estuaries noted above also cause changes in other chemical and physical properties of the sediment, which may influence the animals and plants living there. These changes are: to the water content and interstitial space, with higher water retention in well-sorted fine-grained sediments; to temperature and salinity where they change much more slowly within the sediments than in overlying water or air; and to organic content and oxygen content, where fine-grained sediments are associated with greater organic content, and due to the biological processes within them, with lower oxygen content. The change from an aerated surface sediment layer to deeper anoxic layers takes place closer to the surface in fine-grained muds than in coarse-grained sands. This change, which may be approximated to by measuring the redox potential (E_h), of the sediment, serves to limit the macrofauna in estuarine sediments to species that can form burrows or have other mechanisms to obtain their oxygen from the overlying water.

1.4 Other physico-chemical factors

It is tempting to regard estuarine water simply as diluted seawater. However, there are dangers in this approach, as evidenced by Fig. 1.10. It can be seen that the concentration of bicarbonate ion drops only slightly from the sea to freshwater. The concentration of other ions deviates markedly from the hypothetical dilution line at low salinities (i.e. at less than l0), and certain ions such as phosphate, nitrate, or silicate may even be more abundant in river water than seawater, and thus will show a

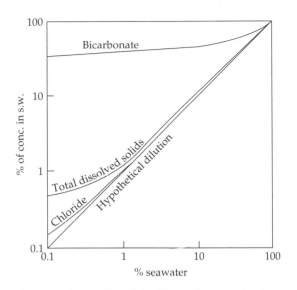

Figure 1.10 The effect of the dilution of seawater by river water on the concentrations of bicarbonate and chloride ions, and the concentration of total dissolved solids. Concentrations expressed as percentage of the concentration in seawater. Note the logarithmic scales. (After Phillips 1972.)

decrease in concentration from rivers to the sea. Dissolved organic matter may also be higher in river water than the sea, and the concentrations of most trace metals are similar in river and seawater and therefore tend to be relatively constant within estuarine water. The distribution of trace metals, such as iron, manganese, cobalt, nickel, copper, zinc, and mainly the distribution and transport of suspended particulate matter, especially the position of the turbidity maximum, control cadmium in estuaries.

Most of the biogeochemical processes involved in the meeting of freshwater and seawater in estuaries occur at very low salinities (<1), where a region known as the freshwater–seawater interface (FSI) has been identified as an important site for chemical and biological reactions. The FSI has also been termed the FBI (freshwater—brackish water interface). It should be noted that changes in the compositional ratios of the major ions such as sodium, and chloride, resulting from mixing river water with seawater are complete at salinities of less than one. These changes are accompanied by dramatic changes in the speciation of metals such as cadmium, with a drastic drop in the proportion of free cadmium ions, and a surge in the concentration of cadmium chloride complexes. The hydrogen ion activity (pH) is at a minimum salinity of 0.5–3, and the drop in pH controls many aquatic reactions, such as dissolution of metal oxides. Thus the addition of sea salt to river water, with its low ionic strength, results in nonlinear chemical perturbations that are amplified in the low salinity waters of the FSI. The chemical changes, which occur at the FSI contribute to the food supply of zooplankton, as is discussed in Chapter 3, and may also be responsible for changes in the composition of the estuarine fauna at low salinities, as discussed in Chapter 2.

The waters flowing into estuaries will convey large quantities of oxygen into the estuary, and additional oxygen will be supplied through the surface of the water and by plant photosynthesis. However, the many organisms living within estuaries, especially in the bottom deposits, rapidly consume the oxygen and thus many sediments are anoxic, except for a thin surface layer. Where excessive organic enrichment occurs the multiplicity of microorganisms so produced may also consume all the oxygen within the water body as well. The temperature regimes within estuaries are often more variable than adjacent waters due to the shallowness of estuaries, which expose the waters to both heating and cooling, and also due to varied inputs of water to the estuary, which may be at different temperatures. A detailed account of estuarine chemistry is beyond this book and here we shall restrict ourselves to the nutrient chemicals that may control the production of plant material, and thus serve as the foundation for the estuarine ecosystem.

Estuaries are generally rich in the nutrients needed for plant growth, especially nitrogen and phosphates, as supplies from rivers, the sea, and the land are continually replenishing the supply of these nutrients. Within the Ems estuary, Netherlands/Germany border, for example, it has been found that the freshwater supplies about 92% of the nitrogen entering the estuary and 71% of the phosphorus. Despite processes within the estuary, mostly burial and denitrification, 60 and 72%, respectively, of the nitrogen are then transported from the estuary to the adjacent North Sea.

The presence of large quantities of suspended sediment in estuaries may cause substantial modifications to the supply of nutrients for plant production. In the Tamar estuary, England, the phosphate supply is the main limitation on the growth of phytoplankton, but the suspended sediments in the water exercise some form of buffering action such that they remove phosphate from the water when the concentration is high and release it when it is low. Thus as the phytoplankton assimilate phosphate from the water, so there will be a tendency for more phosphate to be released from the sediment. The turbidity of estuarine waters may also limit phytoplankton production by restricting the penetration of light.

Studies by Heip *et al.* (1995) have shown that it is the relationship between the benthic and pelagic environment that critically controls the entire functioning of the estuarine ecosystem and they conclude that estuaries are effectively light limited systems where excess nutrient input cannot lead to increase in primary productivity. The overall recycling efficiency of nutrients in estuaries leads to three or fourfold re-use of any nutrients brought into they system. Since the pelagic system is light limited by turbidity, the role of the benthic flora is crucial notwithstanding this recycling of nutrients, there is still a major outflow from estuaries into the sea, and on balance as many nutrients may leave an estuary as enter it.

Nutrient inputs from rivers to the sea are for a variety of reasons, as discussed above, less likely to be "trapped" within the estuaries. So, for example, 75% of the riverine input of nitrogen to the North Sea derives from the Rhine and Elbe rivers, having passed through their respective estuaries. The nitrogen balance of the Westerschelde in the Netherlands shows little cycling within the system. Although denitrification or burial loses 25% of the nitrogen entering the system, most nitrogen is oxidized from ammonia to nitrate and then leaves the system to enter the North Sea. The nitrogen model shows a net annual export to the North Sea of 51,000 tonnes of nitrogen from the Westerschelde, which is about 3% of the total nitrogen input to the North Sea (1.5×10^6 tonnes N year^{-1}), whereas the contribution of freshwater discharge from the Westerschelde is about ten times lower. Soetaert and Herman (1995*a,b*) calculate the nitrogen load of North Sea coastal waters off the mouth of Westerschelde is increased due to tidal flows in and out of the estuary by about 14% per day. Because the estuary mineralizes nearly all allochthonous and *in situ* produced carbon, it is a large net nutrient exporter. Despite the denitrification process that removes part of the nitrogen, the amount of dissolved nitrogen that is flushed to sea is some 25% higher than the amount that enters from rivers or from wastes.

Comparison of nutrient cycling and food webs of six large Dutch estuaries shows that all the estuaries experience increased nutrient loading but, because of turbidity, increased production of primary producers may not necessarily occur. The uptake of dissolved inorganic nitrogen (ammonia and nitrate) in several European estuaries (including the North Sea estuaries of the Ems, Rhine, and Scheldt) has been recently described by Middelburg and Nieuwenhuize (2000), showing that most nitrate entering or produced in estuaries flushes through, and enters the sea, unless denitrification and/or burial in the sediment occurs. Ammonium and particulate nitrogen are efficiently recycled within the estuary.

The distribution of sediments and salinity within estuaries, along with the distribution of oxygen, temperature, and organic debris are all interdependent. The pattern of distribution of these factors varies from estuary to estuary depending on topography and the volumes of water involved. Inherent in the nature of true estuaries around the world are certain ecological common denominators, or factor complexes. They are:

(1) the presence of well-aerated, constantly moving, relatively shallow water, which is mostly free from excessive wave action or rapid currents;
(2) a salinity gradient from near zero up to over 30, and accompanying chemical gradients;
(3) a range of sedimentary particle sizes from colloids to sands and detritus, resulting from land weathering, water transport, and estuarine processes; and
(4) complex molecular interactions in both water and sediments, taking place in an abundance of dissolved and particulate organic matters, microorganisms, and fine sedimentary particles.

Romao (1996) has summarized the habitat of an estuary as

The downstream part of a river valley, subject to tide and extending from the limit of brackish waters. River estuaries are coastal inlets where, unlike 'large shallow inlets and bays' there is generally a substantial fresh water influence. The mixing of fresh water and seawater and the reduced current flows in the shelter of the estuary lead to the deposition of fine sediments, often forming extensive intertidal sand and mudflats. Where the tidal currents are faster than the flood tides mist sediments deposit to form a delta at the mouth of the estuary.

What this and similar descriptions have in common is the recognition that the estuarine environment is a complex and variable environment characterized by salinity, tide, currents, and sediments, which is intimately linked as an ecological unit with its surrounding habitats.

For the biologist looking at the estuarine ecosystem certain characteristic zones emerge, each with typical sediments and salinity. In particular estuaries one zone may occupy a large proportion of the area, and the other zones may be compressed, but a clear sequence can always be seen:

1. *Head*. Where freshwater enters the estuary, and river currents predominate. Tidal but very limited salt penetration. FSI salinity <5. Sediments become finer downstream.
2. *Upper reaches*. Mixing of fresh and saltwater. Minimal currents, especially at high tide, leading to turbidity maxima. Mud deposition. Salinity 5–18.
3. *Middle reaches*. Currents due to tides. Principally mud deposits, but sandier where currents faster. Salinity 18–25.
4. *Lower reaches*. Faster currents due to tides. Principally sand deposits, but muddier where currents weaken. Salinity 25–30.
5. *Mouth*. Strong tidal currents. Clean sand or rocky shores. Salinity similar to adjacent sea (>30).

The relationship of these zones to the distribution of animals and plants will be discussed in Chapter 2.

1.5 Brackish seas and coastal lagoons

So far we have mainly examined the positive estuaries, which are the commonest for those working in the estuarine environment. Mention, however, needs to be made of the range of brackish waters, whose inhabitants face similar problems to those encountered by estuarine organisms. These areas, such as the Baltic Sea, Mediterranean Sea, Black Sea, Caspian Sea, and a variety of coastal and hypersaline lagoons cannot be classified as estuaries, since the definitions of Pritchard and Fairbridge (Section 1.1) emphasize the importance of tides as a distinguishing feature of estuaries and thus explicitly exclude the Baltic Sea and other brackish seas from being designated as estuaries. Indeed we have seen that the origin of the word estuary is from *aestus* meaning tide. Whether it is necessary for all estuaries to be tidal, or whether it is accepted that "*transitional waters*" waters in all areas, irrespective of having tides can be called estuaries is still a matter for debate. To avoid this problem legislators unable to define estuaries are increasingly using the term "*transitional waters.*"

Waters with salinity lower than seawater, and higher than freshwater are called brackish waters (a word derived from the Dutch word *brac*, meaning salty). It should be made clear that the term Brackish waters has a wider

meaning than Estuary. All waters with salinities between sea and freshwater can be called brackish, whether they are large seas, closed lagoons, or tidal estuaries. An estuary is characterized by tidal mixing and flushing as well as a salinity gradient, whereas a brackish water area is characterized principally on the basis of salinity. Thus all estuaries are brackish, but not all brackish waters are estuarine.

The Baltic Sea is a major brackish water sea area, but because of its lack of tides cannot classified as an estuarine. The Baltic Sea (Fig. 1.11) extends from Copenhagen in

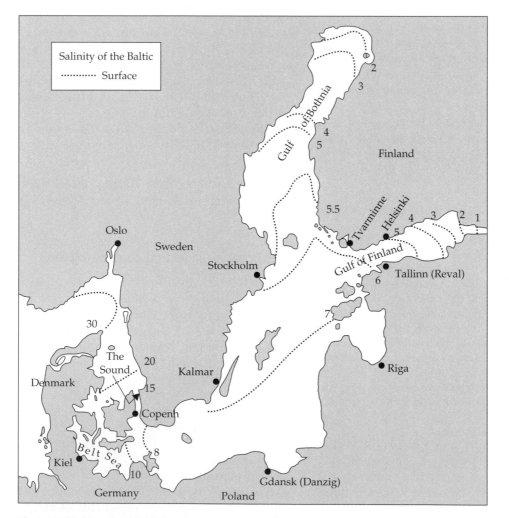

Figure 1.11 Map of the Baltic Sea, showing average surface salinity. (Redrawn from Segerstrale.)

Denmark to Finland, and into the Gulf of Bothnia and the Gulf of Finland reaching St Petersburg. Its maximum depth, off Gotland is >200 m, and it is connected to the North Sea via the shallow Oresund (The Sound) and the Belt Sea. On the seaward side of the Sound the salinity is high (>30). The salinity of the Sound is variable depending on factors such as wind and water level. Saltwater entering the Baltic sinks owing to its greater density and becomes virtually stagnant in the deeper areas of the Baltic. The fresher water from land drainage and snow melt flows outward on the surface. The salinity pattern that develops is relatively stable with a salinity of 8–10 between Denmark and Germany, around 6 between Stockholm and Tallinn, and <2 in the innermost parts of the Gulfs. Unlike estuaries this salinity gradient is relatively stable over hundreds of kilometers, so that the organisms living in the Baltic experience a stable salinity regime coupled with a wide variety of substrates, ranging from rock to soft-sediment. As a consequence many animals are able to penetrate to lower salinities than they do in estuaries, which while it confirms

the ecological master role of salinity also emphasizes that much of the biological patterns seen in estuaries are due to the muddy sediments and the widely varying nature of currents, tides, and associated physical and chemical factors.

Similarly the Mediterranean Sea is an important sea area, which is almost closed except for the Straits of Gibraltar, but cannot be considered estuarine due to its full salinity and lack of tides. Nevertheless some Mediterranean River mouths are commonly called estuaries. The Black Sea was once part of the great Tethys Sea, along with the Sea of Azov, the Caspian Sea, and the Aral Sea (Fig. 1.12). This marine sea contained a typical Mediterranean fauna and during the Miocene era (15–20 million years ago) was split into distinct basins. During later years gradually lowering of salinity occurred, the basins became established as the seas of today and in time their salinity regime became stabilized. The salinity of the Black Sea is typically 17–19. The salinity is uniform and not in a salinity gradient as in the Baltic. The fauna is

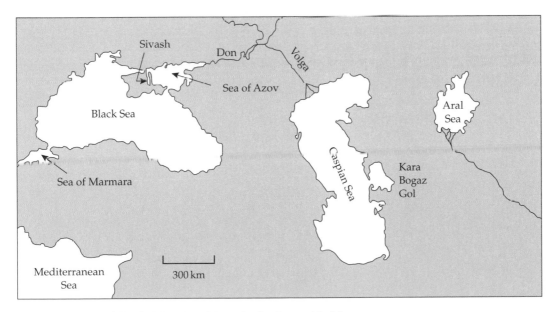

Figure 1.12 Map of the Black Sea, Sea of Azov, Caspian Sea, and Aral Sea.

markedly reduced in diversity compared to the Mediterranean, to which the Black Sea is connected via the Sea of Maramara. Below 180 m, down to the maximum depth of 2243 m there is no dissolved oxygen and free hydrogen sulfide occurs. Thus all but the surface layer is unsuitable for most forms of life. The Sea of Azov is adjacent to the Black Sea and has a salinity of 11. It is not subject to the deoxygenation that afflicts the Black Sea, and the waters are considered productive for fish.

The Caspian and other enclosed seas have brackish waters, with salinities comparable to regions of estuaries, but have neither tides nor any connection with the open sea. The Caspian Sea is the largest land-locked body of water in the world. For centuries it had a salinity of about 12, which was stable as a result of a balance of salt trapped from earlier marine seas, evaporation, and freshwater inflow, mainly from the Volga River. In recent years the freshwater inflow has been drastically reduced by water extraction schemes on the Volga. As a consequence the salinity has increased due to evaporation and the area of the sea has shrunk drastically and catastrophically. The fauna of the Caspian is basically that of the ancient Tethys Sea, which has subsequently evolved in isolation to create a unique relict fauna. The Aral Sea, east of the Caspian has (or had) a salinity of 10, but unlike other brackish seas only 54% of the salinity is due to sodium chloride, and as a consequence the fauna and flora is very limited.

There are two main types of coastal saline lagoons, natural lagoons formed behind shingle or sand barriers, and artificially constructed coastal saline ponds. Lagoons can have salinities from <1 to full strength seawater and, where evaporation occurs, can be hypersaline. Saline lagoons are natural features consisting of shallow open water bodies separated from the sea by a barrier of sand or shingle. Most of the water is retained in the lagoon at low tide, and seawater is exchanged directly through a channel or by percolation through the sediment barrier. Lagoons of an artificial origin can be more common than natural lagoons and are formed as a result of a wide range of man's activities, including sea-wall construction, the digging of borrow pits or of drainage systems in low-lying lands, or as former salt-pans (areas for the preparation of salt by evaporation).

Hypersaline areas are characterized by salinities great than seawater. The Laguna Madre in Texas, for example, experiences salinities ranging from 38 near its entrance to the sea up to 60 or even 80 in areas further from the sea. The increase is mainly due to evaporation exceeding either river or tidal inflow. Just as in brackish areas, which have salinities lower than those of seawater, experience reductions in the number of species, so do hypersaline areas. This phenomenon is discussed further in Chapter 2.

CHAPTER 2

Life in estuaries

2.1 Distribution of estuarine organisms

Estuaries are important to humans and marine life. The tidal sheltered waters of estuaries support unique communities of plants and animals that live at the margin of the sea. The estuarine habitat experiences a steep gradient in salinity conditions, coupled with a high degree of turbidity, which leads to the deposition of muddy intertidal areas. This habitat is rich in wildlife and for many people on this planet is the nearest thing to a natural habitat that they ever see. It is a habitat that man has exploited and often destroyed, but the estuarine habitat remains as one of the most resilient habitats on earth, maintaining their attractiveness for wildlife, despite industrialization and land-claim. It is a habitat that can provide unique ecosystem services to help the sea and mankind, for example, by trapping contaminants in its sediment, while also providing nursery grounds for marine fish and feeding areas for migratory birds.

The life of the plants and animals living in estuaries will be examined in detail in later chapters, but first we need to consider the general features that are common to all of them. All the plants and animals in estuaries originate from either the sea or freshwaters. They either enter estuaries as part of a movement or migrate during their own lifetimes, or else their ancestors entered estuaries and successive generations have become settled within estuaries, so that it is now the only habitat for some species. The species living in estuaries fall into several categories (Box 2.1)

Many studies of the distribution and abundance of animals and plants in estuaries have shown that the number of species within

---

Box 2.1 The ecological categories of estuarine plants and animals

Oligohaline organisms. The majority of animals living in rivers and other freshwaters do not tolerate salinities greater than 0.5, but some, the oligohaline species persist at salinities of up to 5.

True estuarine organisms. These are mostly animals with marine affinities that live in the central parts of estuaries. Most of them could live in the sea but are apparently absent from the sea probably due to competition from other animals. Most commonly at salinities of 5–18.

Euryhaline marine organisms. These constitute the majority of organisms living in estuaries with their distribution ranging from the sea into the central parts of estuaries. Each species has its own minimum salinity that it can live in. Many disappear by 18, but a few can survive at salinities down to 5.

Stenohaline marine organisms. These are marine organisms that only occur in the mouths of estuaries, at salinities down to 25.

Migrants. These animals, mostly fish and crabs, spend only a part of their life in estuaries, with some such as the flounder *(Platichthys flesus)* or the shrimp *(Pandalus montagui)* residing and feeding in estuaries. Others such as the salmon *(Salmo salar)* or the eel *(Anguilla anguilla)* are purely migrants that use estuaries as routes to and from their breeding areas in rivers or the open sea.

estuaries is less than the number of species within either the sea or the freshwaters. Data from a survey of the benthos of the Tay estuary (Scotland), for example, shows a clear decrease in the number of species from the mouth of the estuary inward (Table 2.1). If we also consider the abundance of the various

categories of estuarine organisms a clear pattern emerges as shown, for example, in Tay estuary data (Fig. 2.1).

In the major groups of species living within estuaries a clear generalized pattern of declining diversity can be seen as one enters the estuary from either end. The number of oligohaline species is high in true freshwaters, but declines within estuaries. If the number of species within freshwater is expressed as 100%, then by a salinity of 5 the number of oligohaline species has fallen below 5%. The numbers of stenohaline marine species similarly falls as one enters estuaries from the seaward end, so that of the stenohaline species present in full strength seawater, virtually none are present below a salinity of 25. The numbers of species of euryhaline marine organisms also declines progressively from the sea into estuaries, although in this case some survive down to a salinity of 5. To offset this pattern of declining diversity of species, the numbers of species of true estuarine animals in the sea or freshwaters is low or nil, but increases within estuaries

Table 2.1 Number of benthic species recorded from transects in the Tay estuary

Transect number	Locality	Number of species	Substrate	Salinity
I	Buddon Ness	51	Sand	32–33
II		53	Sand	
III		36	Sand	11–32
IV	Dundee	23	Mud/rock	6–30
V		22	Sand	0.2–20.6
VI		21	Mud/sand	
VII		8	Mud	0–10.7
VIII		7	Mud	
IX	Newburgh	7	Mud/sand	0–0.3

Source: Khayralla and Jones (1975).

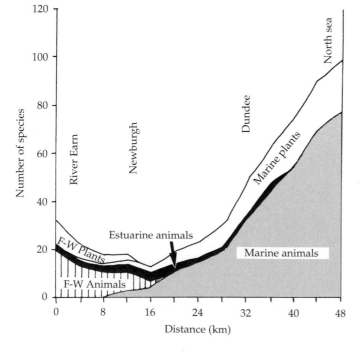

Figure 2.1 Composition of the fauna and flora along the Tay estuary, Scotland. Numbers of species, for each grouping, shown against distance. (After Alexander, Southgate, and Bassindale 1935.)

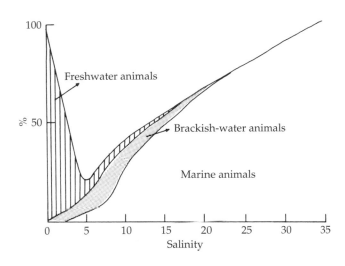

Figure 2.2 Generalized penetration of marine, freshwater, and brackish-water animals into an estuary in relation to salinity. Diversity of marine and freshwater animals shown as a percentage of species diversity in each source habitat. Diversity of brackish-water animals shown as a subdivision of marine animals. (After Remane and Schlieper 1958.)

with a maximum number of estuarine species in the 5 to 18 salinity region. The total number of true estuarine species, as well as the number of migratory species, is, however, relatively low compared to the numbers of oligohaline or marine species. Thus when the numbers of species of all groups living within estuaries is combined in a single diagram, as first drawn by Remane and now known as "Remane's curve" (Fig. 2.2), it can be seen that estuaries are characterized by having fewer species than adjacent aquatic environments.

Salinity determines the distance that a species is capable of penetrating into the estuary, but the full potential of any species to enter the estuary can only express itself when suitable substrates are present. Data from the Avon Heathcote estuary (New Zealand) (Table 2.2), shows how small the estuarine and freshwater component of the fauna is, with only 10 and 8 species each, compared to 146 marine species, of which 72 are stenohaline and 74 are euryhaline. Data from the Potomac estuary (USA) is summarized in Table 2.3, where the sequence of species within the estuary can also clearly be seen.

In most estuaries there is a close connection between salinity distribution and substrate type, with reduced salinity associated with finer substrates. The close connection between

Table 2.2 Faunistic components in the Avon-Heathcote estuary (New Zealand) expressed as species number

	Mouth	Middle	Head	Total
Salinity range	25–34.3	8.4–30.3	0.2–15.7	
Freshwater	0	2	8	8
Estuarine	2	10	8	10
Euryhaline marine	69	67	10	72
Stenohaline marine	74	3	0	74
Total species	145	82	26	164

Source: Knox and Kilner (1973).

Table 2.3 Number of benthic macroinvertebrate species in the Potomac estuary

	Salinity	Number of benthic invertebrate species
Zone		
Tidal fresh	0–0.5	70
Oligohaline	0.5–5	104
Mesohaline	5–18	184
Polyhaline	18–30	199

Source: Lippson *et al.* (1981).

these two physical factors often makes it difficult for the biologist to distinguish their effect. The ecological factors of salinity and substrate are closely interwoven in explaining the distribution of estuarine animals.

By contrast to the decline in the diversity of species within estuaries, it must be emphasized that the abundance of individual species often increases markedly within estuaries. Thus a picture emerges of a greater number of individual animals, but of fewer species. If a comparison is made between the numbers and weight of animals living in the bottom deposits of marine beaches, freshwaters, and estuaries at similar latitudes, it becomes clear that the estuary has the highest number and weight of organisms.

In addition to the physico-chemical factors, discussed in Chapter 1, which control the distribution and abundance of estuarine organisms, Wildish identified three major biotic factors that exercise some control over the numbers of animals inhabiting estuaries: (1) food supply, (2) supply of colonizing larvae, and (3) interspecific competition. The richness of the food supply, especially in the form of detritus, will be discussed later. The supply of larvae within estuaries does present some problems, as planktonic larvae may get carried out of the estuary by flushing currents. Many estuarine animals have thus suppressed the plankton larval stage, and only have bottom-living larval forms. Competition occurs in all ecosystems, but the relatively harsh estuarine environment, which challenges the physiological mechanisms of all species (see Section 2.2) excludes many species, and thus the estuary represents for those species, which can adapt to the environment, an escape from competition in the sea or freshwater. Thus those species that can adapt to estuaries are able to utilize the rich food supply available there, and it is this food supply, which becomes the main factor controlling the biomass and productivity of estuarine animals.

The estuarine habitat is thus not a simple overlapping of factors extended from the sea and from the land, but a unique set of its own physical, chemical, and biological factors. The estuarine habitat has provided the environment for the evolution of some true estuarine organisms, but more importantly has provided a productive environment for those species that have entered it from the river, or from the sea. The estuarine habitat is thus characterized by relatively few species, but these species may be very abundant.

2.2 Problems of life in estuaries

Estuaries are characterized by having abundant populations of animals, but with relatively few species. What features within estuaries serve to restrict the diversity of species, and explain the observed decrease within estuaries? The "stability-time hypothesis" compares the diversity of animals living within different bottom deposits of the sea and estuaries. In this approach the number of species present within a sample is plotted against the number of individual animals present in the same sample. When several environments are compared (Fig. 2.3) a series of curves are produced. For estuarine samples the curve is always nearer the x-axis indicating fewer species in relation to the number of animals present than occurs with marine samples, which confirms the situation for estuaries of low diversity but high abundance. It is suggested that estuarine environments are "physically controlled communities," where physical conditions fluctuate widely and are not rigidly predictable. Thus organisms are exposed to severe physiological stress and/or the environment is of recent past history. By contrast, samples from the deep sea are described as "biologically accommodated communities" with a large number of species per unit number of individuals, where physical conditions are constant and have been uniform for long periods of time. From such data the stability-time hypothesis (as shown in Fig. 2.4) can be developed, with species number diminishing as physiological stress increases.

Two main hypotheses thus emerge to explain the paucity of estuarine species. The first explanation is that of physiological stress, and the second that the estuarine environment is of recent past history and that there has

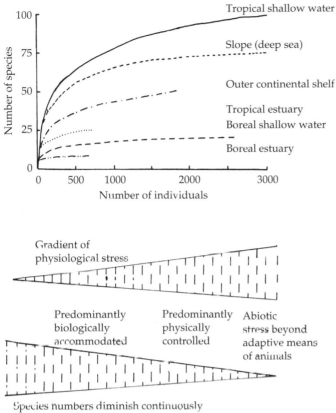

Figure 2.3 Arithmetical plot of the number of species present compared with the total number of individual animals present for sampling stations from different marine and estuarine habitats. (After Sanders 1969.)

Figure 2.4 Representation of the stability-time hypothesis of Sanders showing that as stress increases so the number of species present decreases. As stress increases so the ecosystem comes under the control of predominantly physical factors until it finally becomes abiotic.

been insufficient time for a large number of species to develop within estuaries. These two hypotheses are not mutually exclusive and both could be operative.

The general term "physiological stress" covers a range of problems that confront the estuarine organism. The most conspicuous and probably best-studied physiological factor is salinity, since estuarine organisms are exposed to variable and fluctuating salinities in contrast to the stable salinities, which characterize both marine and freshwater habitats. Other physiological problems, which impinge on the life of an estuarine organism, are the nature of the muddy substrate, both in terms of the fine particulate matter, which can clog delicate organs, and of the virtually anoxic conditions which occur within muds. The responses to

physiological stress can be many and varied, ranging from adaptations of behavioral patterns, and the modifications of particular organ systems, through to the evolution of new races or species adapted to the estuarine ecosystem. These physiological adaptations may take the form of the ability, or otherwise, of an animal to tolerate and thrive in the estuarine environment, such as the tolerance of particular temperature, salinity, or substrate conditions.

The effects of salinity on estuarine organisms are varied. The impact of salinity, as with other environmental factors, is often multivariate. Thus, for example, temperature may interact with salinity, and the response of an animal to a change in salinity may be different at different temperatures. Salinity may affect an animal through changes in several chemical properties of the water. First, the total osmotic concentration of seawater reduces as it is diluted with freshwater. Second, the relative proportion of solutes within estuarine water varies with the salinity, especially at very low salinities as described earlier. Third, the concentration of dissolved gases varies with the salinity, with freshwater containing more oxygen than seawater at the same temperature. Finally, the density and viscosity of water varies according to salinity, with freshwater lighter than saltwater. Thus when it is reported that an animal responds to a change in salinity it is important to check whether the animal is responding to the change in salt concentration, or to any of the above factors instead. The effects of salinity on an animal may also differ at different stages in the life cycle. In general it appears that animals are more sensitive to extremes of salinity at the egg stage, when newly hatched, or when in adult breeding condition, than they are at intermediate stages of growth.

Salinity stress may invoke several responses. Confronted by an abnormal salinity an animal may seek to escape or otherwise reduce contact with the water. Such behavioral responses are common in many estuarine animals. For actively swimming animals, such as fish, it is relatively easy to escape from an adverse salinity, but sedentary animals such as barnacles can only respond by reducing contact and attempt to seal themselves inside their shell. For a burrowing animal it is often possible to retreat into the burrow, or digs deeper, when confronted by an abnormal salinity, but such a solution is only useful if the abnormal salinity is of a short duration. If it is not possible to escape or reduce contact with the abnormal salinity, then the organism must rely on physiological responses. Either it can allow the internal environment (blood, cells, etc.) to become osmotically similar to the external environment (sea or estuarine water); or else it can attempt to maintain the internal environment at an osmotic concentration different from the external environment by the process of osmoregulation. For most stenohaline marine animals the former situation applies. When living in the sea their internal environment is isosmotic (osmotically similar) with seawater, thus although the ionic composition of the internal environment may be different from seawater, the total salt concentration is similar to the seawater. As such an animal enters an estuary the osmotic concentration of the internal environment falls to maintain equality with the less saline estuarine water. This passive tolerance of osmotic change is, however, only possible down to a salinity of 10–12 because below this salinity the cells and tissues of the animal may cease to function.

As an alternative to the passive tolerance of osmotic dilution, an animal may attempt to osmoregulate and maintain the internal environment at a different osmotic concentration from the outside. Thus, for example, an oligohaline, or true estuarine animal, which is living in salinities below 10, must keep the internal concentration at a level greater than the outside. Such an animal, termed a hyperosmotic regulator, will strive to maintain its blood concentration greater than a salinity of 10–12, even though the external environment is lower, and thus enable the cells and tissues to function.

An animal, which hyperosmoregulates when living in low salinities, is faced with two options when living at high salinities. Either it can become isosmotic with the seawater, or it can seek to minimize the fluctuations of the external salinity regime by maintaining the blood less concentrated, or hypo-osmotic, than the external salinity. In the latter case the animal may thus live in salinities ranging from near zero to over 35, while the blood concentration may only rise from about 12 to 22 salinity. Such an animal, termed a hyper-hypo-osmoregulator can thus achieve a much more stable internal environment than is the case with a hyper-iso-osmoregulator, which has still achieved greater stability than the completely iso-osmoregulator. These patterns of osmoregulation are presented in Fig. 2.5.

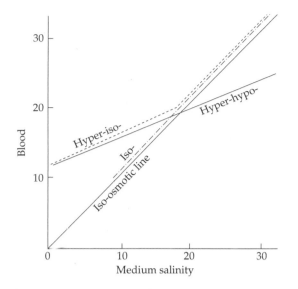

Figure 2.5 Typical patterns of osmoregulation. Iso-osmotic species are typically marine species. Typical hyper-iso-osmotic species are *Nereis, Corophium*, or *Gammarus*, and typical hyper-hypo-osmotic species are *Crangon, Praunus*, or *Neomysis*. The concentrations of the blood and the water medium are expressed as practical salinity units.

The most common feature of the physiology of estuarine animals is thus their ability to tolerate variations in their environment. With regard to salinity they are generally *euryhaline*, tolerating a range of salinities, in contrast to the *stenohaline* marine and freshwater animals, which can tolerate only a small range of salinities. The ecophysiology of estuarine animals can in general be termed *euryoecious*, that is, living in a wide range of ecological conditions. It should always be remembered that only few of the many animals inhabiting the aquatic world have developed these abilities, and among those that have these abilities the pressures of competition cause different species to live in different parts of the estuary.

Estuaries offer excellent opportunities for the study of competition, coexistence, and evolution of closely related species. The presence of closely related species in European estuaries is generally attributed to the brief geological history of present-day estuaries. Estuaries are constantly changing due to the deposition of sediments, and to changes in sea level following the last Ice Age. Thus, the estuaries that we see today in Europe and North America did not exist 10,000 years ago, and might be unrecognizable if we were able to visit them in 10,000 years time. Estuaries as environments have been present for many millions of years, but not those that most of us currently see and know. It has been suggested that the life history of individual estuaries is too short for many species to colonize and adapt to them, and this has been given as a reason for the relatively few species in them today. Further, the environmental gradients that characterize estuaries may enable us to examine what factors cause one species to be ecologically and genetically separated from another.

Within the polychaetes there are three species of the rag-worm *Nereis* (also known as *Hediste*) present in European estuaries, namely *Nereis diversicolor, N. virens*, and *N. succinea. N. virens* is larger and more voracious than the others are, but intolerant of low salinities. *N. diversicolor* can live in a wide range of salinities, but is apparently excluded from higher salinities by competition from *N. virens. N. succinea* forms an intermediate species, but may disappear completely with low temperatures. Four species of the car-worm *Nephtys* occur in the estuaries, namely *Nephtys caeca, N. cirrosa, N. hombergi*, and *N. longosetosa*. These species inhabit at least three separate ecological niches. *N. caeca* occurs subtidally in muddy sands, while *N. hombergi* lives both intertidally and subtidally in even muddier substrates. *N. cirrosa* and *N. longosetosa* both inhabit clean sands. Thus in these species it can be seen that substrate preference acts as the main factor isolating one species from another.

There are four similar Hybrodiid snails in estuaries. *Potamopyrgus jenkinsi* is found typically in freshwater and in estuaries with salinities up to 15. *Hydrobia ventrosa* is found between 6 and 20, *H. neglecta* from 10 to 24, and *H. ulvae* from 10 to 35. Salinity is not a complete explanation, as *H. ventrosa* and *H. neglecta* occur mainly in areas of stable salinities such as the

Baltic Sea and the Belt Sea, whereas *H. ulvae* dominates in the fluctuating salinities that typify many estuaries. Two species of cockle occur in European estuaries, *Cerastoderma* (=*Cardium*) *edule* and *Cerastoderma glaucum* (=*C. lamarcki*). In Danish estuaries *C. glaucum* is the principal cockle at salinities below 18, while both species occur at salinities of 20–25. At 30–35, *C. edule* is the principal species, but at salinities of 60 in the Sea of Azov *C. glaucum* occurs. It thus appears that the "brackish-water cockle" *C. glaucum* can tolerate a range of salinities wider than *C. edule*, but in normal seawater the "edible cockle" *C. edule* is the dominant species.

The two main *Corophium* species in estuaries appear to be separated by both substrate and salinity. *Corophium volutator* is more tolerant of low salinities, while *Corophium arenarium* is more abundant at high salinities and with sandier substrates. The importance of osmo-regulation as a means of limiting the penetration of species into estuaries has been shown in the isopods of *Jaera albifrons* group. *Jaera ischiosetosa* can survive for several days in freshwater and is able to maintain a steady internal osmotic pressure even in very dilute seawater, and is the species, which penetrates furthest into estuaries. *Jaera forsmanni* and *Jaera praehirsuta* have lower survival rates in dilute salinities and are less capable of maintaining a steady state in dilute seawater, and are found at the seaward end of estuaries. *J. albifrons* is intermediate in both osmoregulatory ability and in its ecological distribution. These physiological differences coupled with the observed ecological difference show how genetic isolation between the species can occur.

In considering the possible causes for the brackish water minimum of species at a salinity of 5, the so-called Remane's minimum shown in Fig. 2.2, Wolff (1973) and Deaton and Greenberg (1986) reject pollution, size of estuaries, turbidity, or the critical salinity hypothesis of Khlebovich, as satisfactory explanations. Pollution is rejected as an explanation because the decrease in species within temperate estuaries occurs in both unpolluted and polluted estuaries alike, although the decrease will be more marked in polluted situations (Chapter 6). The size of estuaries as a reason is rejected because the decrease in species occurs in both small coastal estuaries as well as large brackish seas, such as the Baltic. Furthermore, small freshwater bodies have greater species diversity than larger estuaries nearby. The turbidity of estuarine waters may inhibit marine species with delicate gill structures, but the decline of species in brackish waters occurs in both turbid estuaries, as well as estuaries with lower sediment loads, and non-turbid brackish seas. The critical salinity hypothesis of Khlebovich (1968) suggested that for many purposes estuarine and brackish waters with more than a salinity of 5 may be regarded as dilutions of seawater maintaining a constant ionic ratio, but that below a salinity of 5 many chemical properties of estuarine water change. Accordingly he suggested that these chemical peculiarities of estuarine water are responsible for the observed eco-physiological barrier at a salinity of 5 dividing the marine and freshwater components of the estuarine fauna. This hypothesis is, however, invalidated by many important estuarine animals, which live across the 5 salinity divide, by the fact that the reduction in species occurs progressively all the way from the mouth of an estuary, and by the fact that the major ionic perturbations occur at very low salinities, the FSI at <1, at which salinity species diversity is increased compared to 5.

If the above reasons are rejected as unsatisfactory explanations for the reduction of species in brackish and estuarine waters, what can be suggested? The main hypothesis that remains is that of physiological stress. The salinity conditions of estuaries present, as we have already seen, many challenges to the physiological processes of the estuarine habitants. The main challenge of living in a range of salinities rather than the constant salinities of the sea or freshwaters is met by the range of adaptations both genetic and behavioral that occur in brackish-water species. It appears that few of the many animals living in the sea have these osmoregulatory abilities, and the decline of species within estuaries can be seen as a progression of species each reaching their own physiological limit.

Studies of the fauna of brackish seas, such as the Baltic, where there is a progressive decrease in salinity, reveal an accompanying decrease in species diversity just as seen in typical estuaries. The rate of decrease of species is, however, generally less marked in brackish seas than in estuaries. This difference in the rate of decrease may be explained by the stability of the salinities in each area. In the Baltic it is found that more species are able to live at lower salinities in the stable low salinities of the Baltic than they are able to in the salinity regimes of estuaries, which vary with each tide as well as seasonally. It thus appears that the fluctuations of salinity within estuaries enhances the rate of decrease of species, but since a decrease occurs in both stable as well as fluctuating situations instability cannot be an exclusive explanation. A further factor that permits survival of fauna at lower salinities in brackish seas than in estuaries is that low salinity areas in estuaries are invariably associated with high turbidity and fine sediments. Whereas in brackish seas the waters are less turbid due to their lack of tides and there may often be a wider variety of habitats ranging from rocky shores through to soft-sediments.

2.3 The estuarine food web

The estuarine food web is dependent on the input of energy from sunlight and the transportation of organic matter into the estuary by rivers and tides. Within the estuary the plants, or primary producers, convert these inputs into living biological material. As the plants grow they are consumed by the herbivores, or primary consumers, which are in turn consumed by the carnivores, or secondary consumers. In order to understand the conversion of material as it passes through the links of the estuarine food web we need to study and quantify the life of each member of the food web.

The simplest means of quantification is to identify, count, and weigh the organisms within particular parts of the estuary. The identification of the organisms is of course the first step, and can generally be performed with the aid of guides to the seashore. The estimation of the number of organisms depends on choosing a unit area, usually 1 m^2, and counting the animals and plants within the area. For small organisms it is usually impracticable to count everything in 1 m^2, so one would count those within a smaller area such as 10 cm^2 and multiply the result by an appropriate factor to produce a result in terms of 1 m^2. Such counts need to be repeated several times and the mean result determined. If the animals are buried in the sediment, as will be the case in many estuarine species, then the sediment must be sieved through an appropriate sieve to extract the animals. A 1-mm sieve is suitable for bivalve molluscs, but for most other estuarine animals a 0.25 or 0.5 mm (-250–$500 \mu m$) sieve is needed to be certain that a true estimation of abundance is obtained.

The typical results of an exercise in counting estuarine animals is shown in Fig. 2.6, from which a clear pyramid of numbers can be seen, with many small animals, decreasing to fewer larger animals. If, however, the animals are weighed the reverse picture can be seen, Fig. 2.7, with the total weight of the smallest animals considerably less than the total weight of the larger animals, the pyramid of weight. The name given to the weight of the organisms is the *biomass*. The measurement of biomass alone, important though it may be in comparing the immediately available standing crop from one part of an estuary to another part, is quite inadequate for the purpose of estimating the amount of food, which one trophic level, such as the plants, can make available to the animals feeding on them. In order to understand these feeding, or trophic, relationships we need to know the rate of production of organic matter by the various members of the estuarine

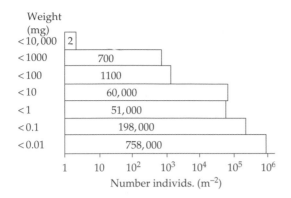

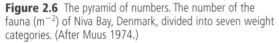

Figure 2.6 The pyramid of numbers. The number of the fauna (m^{-2}) of Niva Bay, Denmark, divided into seven weight categories. (After Muus 1974.)

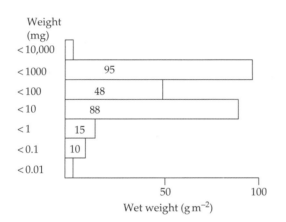

Figure 2.7 The pyramid of weight. The weight of the fauna (m^{-2}) of Niva Bay, Denmark, divided into seven weight categories. Compare with Fig. 2.6. (After Muus 1974.)

Table 2.4 Average caloric values

Taxon	Ash-free dry weight		Number of species
	kcal g^{-1}	kJ g^{-1}	
Sponges	5.79	24.2	34
Jelly fish, Sea anemones	4.67	19.6	20
Annelids	5.32	22.3	26
Bivalve molluscs	5.30	22.2	115
Other molluscs	5.51	23.1	143
Crustaceans	5.35	22.4	204

Source: Beukema (1997).

Table 2.5 Energy budget for intertidal *Scrobicularia plana*

Component	Abbreviation	Value
Mean annual biomass	B	203 kcal m^{-2}
Change in biomass	ΔB	40 kcal m^{-2} year^{-1}
Mortality	E	20 kcal m^{-2} year^{-1}
Production	P	124 kca1 m^{-2} year^{-1}
Assimilation	A	600 kcal m^{-2} year^{-1}
Gamete production	G	64 kcal m^{-2} year^{-1}
Respiration	R	476 kcal m^{-2} year^{-1}
Egesta	F	388 kcal m^{-2} year^{-1}
Consumption $(A + F)$	C	988 kcal m^{-2} year^{-1}
Production/biomass ratio	P: B ratio	0.61
Assimilation efficiency	$A \times 100/C$	61%
Net growth efficiency	$P \times 100/A$	21%
Ecological efficiency	$E \times 100/C$	2%

Source: Hughes (1970).

ecosystem. The distinction between the rate of production (or productivity) of organic matter by an organism or community, and the biomass (or standing crop) consisting of the organism itself or the community, is fundamental.

A summary of the terms and units used in productivity studies is given in Box 2.2. The derivation of various units of biomass and its conversion into energy units is given in Box 2.3. A summary of approximate conversions from one unit to another is given in Box 2.4 and examples are given in Table 2.4.

The production of a population may be derived by one of two approaches. Either all the growth increments of all the members of the population under study are added for 1 year, or else growth is ignored or production is derived from the addition of the change in the biomass over 1 year plus the mortality due to predation etc. over the same period. An energy budget for an intertidal population of the bivalve *Scrobicularia plana* may be taken as an example of the utilization of energy by an estuarine invertebrate (Table 2.5). Note that

Box 2.2 Terms used in productivity studies, which are utilized in this book

Term	Abbreviation	Definition
Biomass	B	Amount of living substance constituting the organism under study
Mean biomass	B^-	Mean amount of living substance constituting the organism during the time of study (typically 1 year)
Consumption	C	Total intake of food or energy
Egesta	F	That part of the consumption that is not absorbed but is voided as feces
Absorption	Ab	That part of the consumed energy that is not voided as feces
Excreta	U	Absorbed consumption that is passed out of the body as secreted material, for example, urine, mucus
Assimilation	A	That part of the consumption that is used for physiological purposes (C excluding U and F)
Production	P	That part of the assimilated energy that is retained and incorporated in the biomass of the organism. May be regarded as "growth"
Respiration	R	That part of the assimilated energy that is converted into heat directly or through work
Gonad output	G	Energy released as reproductive bodies
Yield	Y	That part of the production utilized by another organism

Note: Assuming the conservation of energy we may derive the following equations:

Consumption $C = P + R + G + U + F$ = (total energy budget)
Absorption $Ab = C - F = P + R + G + U$
Assimilation $A = C - (U + F) = P + R + G$

Derived from the above is the $P:B$ ratio, which is the ratio of the production of an organism to the mean biomass of that organism over 1 year.
Source: Crisp (1984).

79% of the energy assimilated was utilized in respiration, which is a typical value for energetics studies, and only 2% of the energy consumed by *Scrobicularia* was consumed by the next trophic level.

The relationship between production and mean annual biomass is known as the turnover ratio, or $P:B$ ratio. Comparison of various temperate benthic invertebrates shows that the $P:B$ ratio varies only between 2.5 and 5, with a mode of about 2.5. This value shows that production over a study year is 2.5 times greater than the mean biomass, and emphasizes that production is a much more important measure of the transfer of energy within an ecosystem than is the biomass. The $P:B$ of estuarine animals from temperate areas may vary from less than 1, to well over 5. The significance of this ratio is that it is a clear indication of the ecological performance of a population. Two populations may, for example, have similar biomasses, but different $P:B$ ratios, thus the one with the high ratio will produce more organic material to be made available to predators than the one with the lower ratio. In comparing different animals at one site, smaller animals will have higher $P:B$ ratios than large animals.

Box 2.3 Units used in productivity studies, which are utilized in this book (After Crisp 1984)

Term	Abbreviation	Explanation
Abundance	Number m^{-2}	The number of organisms contained within 1 m^2 either on the surface or in the substrate or water column underlying 1 m^2
Biomass (1)	g total wt	The total weight in grams of the living wet organisms
Biomass (2)	g dry wt	The weight of organisms after drying at 60–100 °C (approx. 17–7% of (1))
Biomass (3)	g flesh dry wt	The dry weight of living animal, which remains after a nonliving shell has been removed (applicable to hermit crabs, bivalve molluscs, etc.)
Biomass (4)	g ash-free dry wt	The weight of an organism which has had a nonliving shell removed, the water evaporated off and the ash weight of inorganic matter determined by ignition of the sample at about 450–600 °C. The result is regarded as the best measure of the weight of living material *Biomass may be expressed as energy units:*
Calories	cal	*Definition:* Quantity of heat required to raise the temperature of 1 g of water through one centigrade degree at 15 °C *Application:* The calorific content of a biological sample can be determined in a bomb calorimeter and represents the energy released during complete combustion of 1 g of sample
Joule	J	SI unit, which has replaced the calorie 1 cal = 4.184 J. The results of energy units are usually multiplied 1000 fold and expressed as kilocalorie or kilojoules per gram (or kcal g^{-1})
Gram carbon	gC	The carbon content of a sample can be determined and the biomass expressed in terms of carbon weight
Gram nitrogen	gN	The nitrogen content of a sample can be determined and the biomass expressed in terms of nitrogen weight

Biomasses expressed as gram carbon or gram nitrogen, like gram ash-free dry weight, are intended as measures of the living tissue present in a sample, and provide a useful alternative to the determination of the energy content of the material.

The above biomasses (1–4) can also be expressed in kilogram or milligram. However, the values derived take no account of the energetic value of the tissue, which may vary from species to species.

Reasonable estimates of annual production and its distribution among size groups in benthic communities can be derived from knowledge of biomass and the size structure of the community (Fig. 2.8). As a consequence of many productivity studies, the $P:B$ ratio has been extensively investigated over the past 30 years. Our understanding has passed from a realization that body size and $P:B$ are linked to one where the relationship between the two

---

Box 2.4 Conversion of units to and from each other. Conversions of units of biomass into units of energy vary from species to species, and from season to season, however, certain approximations can be used as indicated below (Data from various sources)

1 gC approx. = 10 kcal
1 gC approx. = 2 g ash-free dry wt
1 g ash-free wt approx. = 21 kJ
1 g organic C approx. = 42 kJ
1 litre oxygen = 4.825 kcal (oxycalorific equivalent)
1 g carbohydrate approx. = 4.1 kcal
1 g protein approx. = 5.65 kcal
1 g fat approx. = 9.45 kcal
1 cal = 4.184 J

Some selected examples:
1 g flesh dry wt *C. edule* = 18.57–19.46 kJ
1 g ash-free dry wt *C. edule* = 19.99–21.50 kJ
1 g flesh dry wt *C. volutator* = 13.26–16.82 kJ
1 g ash-free wt *C. volutator* = 18.66–20.88 kJ
1 g flesh dry wt *M. edulis* = 19.95–20.83 kJ
1 g ash-free dry wt *M. edulis* = 21.71–22.51 kJ
1 g flesh dry wt *M. balthica* = 16.10–19.04 kJ
1 g ash-free dry wt *M. balthica* = 19.38–21.46 kJ
1 g flesh dry wt *N. diversicolor* = 16.44–19.70 kJ
1 g ash-free dry wt *N. diversicolor* = 20.87–22.42 kJ

Thus 1 g flesh dry wt estuarine invertebrate approx. = 18 kJ
Thus 1 g ash-free dry wt estuarine invertebrate approx. = 21 kJ

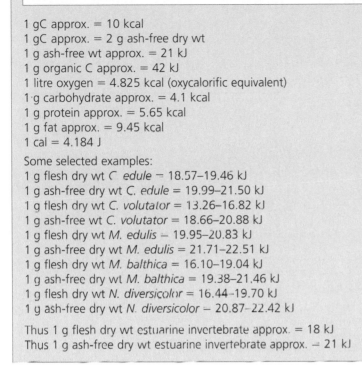

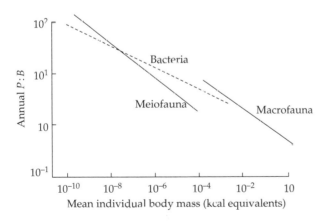

Figure 2.8 Relations between annual $P:B$ (production:biomass ratio) and mean individual body mass (as kilocalorie equivalents) for benthic organisms. Geometrical mean functional regressions are shown for bacteria, meiofauna, and macrofauna. (After Schwinghamer *et al.* 1986.)

is routinely accepted for productivity calculations. The most commonly used relationships are shown in Table 2.6. While they work well in temperate situations, there are some discrepancies when used in tropical situations. Similar relationships can be shown between the annual respiration (R) and annual productivity (P) (Table 2.7).

Table 2.6 Productivity as a function of biomass (B), body size (M, kcal equivalents) and temperature (T, °C)

Group	Relationship
Bacteria	$P:B = 0.696*M^{-0.208}$
Meiofauna	$P:B = 0.73*M^{-0.337}$
Macrofauna	$P:B = 0.525*M^{-0.304}$
Benthic invertebrates	$P:B = 0.65*M^{-0.37}$
Benthic invertebrates	$Log_{10}P = -0.473 + 1.007*log_{10}B - 0.274*log_{10}M$
Benthic invertebrates	$P = 0.0049*B^{0.80}T^{0.89}$

Source: Wilson (2002), and references therein.

Table 2.7 Productivity (P) as a function of respiration (R)

Group	Units	Relationship
Poikilotherms	kcal	$P = 0.64*R^{0.85}$
Macrofauna	kcal	$Log_{10}R = 0.367 + 0.993*log_{10}P$
Estuarine fauna	gC m^{-2} year^{-1}	$Log_{10}R = 0.081 + 1.02*log_{10}P$
Estuarine Birds	kcal	$Log_{10}P = 0.79*log_{10}R - 1.055$

Source: Wilson (2002), and references therein.

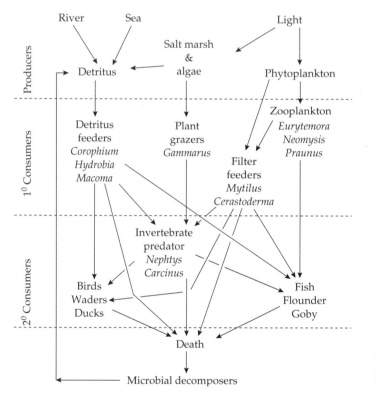

Figure 2.9 An estuarine food web. The arrows indicate the feeding relationships of a typical northern hemisphere estuary. The dotted lines indicate the division into producers, primary consumers, and secondary consumers used in this book.

Organisms within the estuarine ecosystem may be assigned to trophic (or feeding) levels following the trophic—dynamic approach to ecology. The ecosystem may be defined formally as the system composed of physical–chemical and biological processes active within a space–time unit of any magnitude, and thus includes the biotic community of organisms plus its abiotic environment. Within the biotic community we can assign organisms to trophic levels, which are the groups of organisms, which share a common method of obtaining their energy supply. The first trophic level is the producers, which are the plants obtaining their energy by photosynthesis. Following them are the primary consumers, which are herbivorous animals feeding on plant material. Following these are the secondary consumers, which are carnivorous animals feeding on the primary consumers. Eventually we reach the tertiary consumers feeding on the secondary consumers. At each stage in this trophic sequence energy is consumed, and while some of it is rejected as waste, or converted into bodily growth, the majority of the energy is dissipated as heat following respiration. The loss of energy due to respiration is progressively greater for higher levels in the trophic sequence, as is the efficiency of utilization of the food supply. Due to the losses of energy in food chains there are rarely more than five trophic levels. Apart from the loss of energy from each trophic level due to respiration and predation, other materials such as excreta or dead organisms pass to the decomposer trophic level where microorganisms break down the material until it too is lost as heat or else reutilized as detritus by the detritivores, which are regarded as members of the primary consumer level.

This model of trophic transfer is essentially a food chain with one level linking directly to the next. In practice, however, when examining ecosystems we are looking at a food web with some consumer animals, for example, feeding on prey from different trophic levels. In the succeeding chapters of this book we shall examine the estuarine ecosystem as a series of trophic levels, but the reader should be cautioned that it is not always easy to assign a particular organism to a particular level. In Fig. 2.9 an attempt is made to show some of the links within a typical north European estuary. Such a food web is derived from an understanding of the biology and feeding habits of the various members of the estuarine ecosystem. The studies of ecosystem processes and the factors controlling energy flow, either at the species or the community level are the basic building blocks of our understanding of the estuarine ecosystem. Initially many studies have concentrated on single species, measuring their consumption, growth, and the production of energy. In time such studies can be combined in order to build up food webs or trophic diagrams, and eventually we can measure the trophic structure, cycling, and the ecosystem properties of estuaries.

Throughout this book we shall attempt to explain both the energy sources, which support a particular organism, as well as the contribution that organisms make to the functioning of the estuarine ecosystem of which it is a member. We shall consider the ecology of each trophic level in turn, before considering how the systems operate together.

CHAPTER 3

Primary producers: plant production and its availability

3.1 Introduction

Within the estuarine ecosystem there may be several sources of plant production. Growing on the intertidal zones are usually a number of salt marsh plants. In most European estuaries the salt marsh plants are confined to the topmost part of the intertidal zone where they are not covered by the tide every day, but in many American estuaries the salt marsh plants may occupy the major part of the intertidal area and be immersed at each tide. In other parts of intertidal zone may often be found the eel-grass (*Zostera*), which is a true flowering plant, or representatives of the algae. Some of the algae are attached to rocky outcrops such as the typical seaweeds, for example, *Fucus* species. Also growing directly on the surface of the mudflats may be the filamentous algae, *Enteromorpha* species, or the single-celled microphytobenthos (also known as epibenthic algae). Within the water body are found floating members of the phytoplankton, for example, diatoms or dinoflagellates.

The production of all these various plants is of course dependent on both sunlight and temperature, and may also be potentially limited by the availability of nutrients, especially nitrogen and phosphorus. These nutrients are typically rich in estuarine waters, having been carried there from the sea, rivers, or land adjacent to the estuary. Within the estuary the nutrients are utilized by the plants, and following the death of the plant become recycled by the processes of decomposition to be utilized again by the plants. High levels of primary production occur in estuaries in comparison to the open sea or coastal waters, due mainly to the high nutrient levels in estuaries (Table 3.1).

In considering the role of the primary producers as food sources for the primary consumers of the estuarine ecosystem it is necessary to consider the importance of detritus. Detritus has been defined as "all types of biogenic material in various stages of microbial decomposition, which represents a potential energy source for consumer species." Much of this biogenic material is fragments of plant material. Strictly speaking the bacteria and other microbial organisms, which live on and decompose the plant fragments are a second trophic level, dependent on the first trophic level, the plants. The phytoplankton, benthic microalgae, plant fragments and their decomposers, however, become so intertwined, that the food for the primary consumer animals is

Table 3.1 Net primary production in various marine habitats

	Open sea	Coastal zone	Upwelling regions	Estuaries (and coral reefs)
% Sea	90	9.4	0.1	0.5
Net primary production ($gC\ m^{-2}\ year^{-1}$)	50	100	300	1000
Net primary production ($10^9\ tonC\ year^{-1}$)	16.3	3.6	0.1	2.0

Source: Strachal and Ganning (1977).

generally called particulate organic matter (POM), without regard to its exact origin.

In this chapter, therefore, we shall examine both the primary production of the salt marshes and algae (benthic or planktonic), and the limitations placed upon this productivity by nutrient availability. Also the fate of the plant material as it is fragmented and decomposed, and thereby becomes more available to consumer animals. The term primary production is generally viewed as the assimilation of inorganic carbon and nutrients into organic matter by autotrophs. Hence primary production is a rate. Although this definition also includes production by chemoautotrophs, this is not normally measured, because most primary production measurements on phytoplankton (and other aquatic plants) are made with the ^{14}C method, and with this method the dark-bottle measurements are usually subtracted from the light-bottle values to obtain a true photosynthesis rate. Community ecologists generally use the term "gross primary production" as organic carbon production by the reduction of CO_2 as a consequence of photosynthesis. Net primary production is then defined as gross primary production minus autotrophic respiration. Within the literature there may, however, be confusion regarding these terms due, in large part, to the wide variety of techniques in use and the reader is referred to Underwood and Kromkamp (1999) for fuller details of definitions and techniques used to measure primary production in estuaries.

Estuaries are heterotrophic systems, in which consumption exceeds production, and it is the overall organic loading that controls primary and secondary production (Heip *et al.* 1995). In terms of loading, estuaries receive large quantities of allochthonous inputs, that is organic matter generated outside the system and transported into the estuary where it is then available for heterotrophic consumption. There are three principal sources of allochthonous input, namely tidal import from the sea, riverine sources and sewage and waste disposal. Tidal imports vary with the size of the tide and can be linked directly to the volume of water exchanged on each tide, but these have to be balanced against the corresponding export of material on the ebb. It is this balance, which determines whether or not tidal movements function as a net import (flood dominated) or export (ebb dominated) of material.

The concept of ebb or flood dominated tidal flows and interest in the transport of materials in estuaries resulted in the "Outwelling Hypothesis" of Odum (1968). This stated that marsh–estuarine ecosystems produce more organic material than can be utilized or stored within the system and that the excess material is exported to the coastal ocean where it supports near coastal ocean productivity. Since this was formulated several investigations of material transport have been carried out. It has been found in general that in European systems there is usually a net import of particulate materials while in North American estuaries, export of material is normally observed (Table 3.2).

Table 3.2 A summary of material (particulate and dissolved) transport in estuarine systems

Continent	Number of systems studied	Form	Net import	Net export	Both import and export
United States	16	Particulate	2	12	1
		Dissolved	1	7	5
Europe	10	Particulate	9	0	1
		Dissolved	3	2	1

Source: Wilson (2002).

3.2 Salt marshes

The estuaries of the southeastern coast of America are dominated by large stands of the marsh grass, *Spartina*, especially *Spartina alterniflora*, which may occupy up to 90% of the intertidal area. These salt marshes have long been recognized as being among the most productive ecosystems in the world. Maximum production (up to 3300 g dry wt m^{-2} $year^{-1}$ of above-ground material) occurs in southern US states, and this decreases northwards. The high overall levels of production are attributed to the ample supply of dissolved nutrients, coupled with a long growing season and hybrid vigour displayed by the *Spartina* plants. While the net production of *Spartina* is generally high, the levels reported even from one latitude are rather variable. A major factor in this variability is the tidal range with the net production increasing as the tidal range increases, due apparently to increased availability of the nutrient nitrogen. While *Spartina*-dominated salt marsh estuaries certainly support coastal ecosystems through their exceedingly high productivity and the subsequent export of detritus, many of the results and conclusions are, however, as varied as the sites selected for study. Great care should be exercised in applying the results from one estuary to another, which may have different current patterns and topography.

Estuarine salt marshes are highly productive ecosystems, with gross primary production rates varying from 100 to 1000 s gC m^{-2} $year^{-1}$. Despite this high potential input of carbon from marshes, the role of estuarine marshes as a source of particulate organic matter for the estuary proper is variable. The majority of the carbon fixed is consumed by respiration, and only a fraction of the gross primary production, namely the net ecosystem production accumulates in the marsh ecosystem or becomes available for export to adjacent waters. Estimates for the export (or import) of energy from American Atlantic and Gulf Coast salt marshes (Table 3.3) show that most marshes export significant amounts of carbon to adjacent waters. The outwelling of organic carbon from salt marshes in the United States is attributed to the presence of *S. alterniflora* in the lower intertidal zone. European Atlantic salt marshes are generally confined to the uppermost part of the intertidal and there are no indications that the European marshes export significant amounts of particulate organic carbon.

Teal's classical study of energy flow in a salt marsh ecosystem in Georgia was one of the

Table 3.3 Production, respiration, net ecosystem production, burial, and exchange of American salt marsh systems (gC m^{-2} $year^{-1}$)

Marsh system	Primary production (*P*)	Respiration (*R*)	Net ecosystem production (*P−R*)	Burial (*B*)	Export: mass balance (*P−R*)−*B*	Export: direct flux
Sippewisset, MA	1918	1678	240	89	151	54–76
Flax Pond, NY	1725	1425	300	200	100	80–73
Duplin River, GA	3941	2859	1082	29	1053	379
Barataria, LA	4261	4060	201	174	27	20
Bissel Pond, RI	960	980	−20	0	−20	NR

Note: The mass balance export requires extrapolation to the entire salt marsh, and direct flux is based on samples taken directly from tidal creeks.

Source: Heip *et al.* 1995 and references therein.

first studies to present a complete energy flow for any ecosystem, and he showed that the salt marsh under study received 600,000 kcal m^{-2} year^{-1} of sunlight, of which 8295 kcal m^{-2} year^{-1} became net primary production within the salt marsh. Twenty percent of this net primary production was due to benthic algae, with 80% of the net primary production due to *Spartina* grass. The algae were utilized by consumer animals directly, but most of the *Spartina* became detritus and was subject to decomposition by bacteria, with much of the *Spartina* production dissipated as bacterial respiration (Fig. 3.1). A small amount of the

Spartina production was also assimilated directly by herbivorous insects.

Nitrogen is a key nutrient in the productivity of coastal ecosystems, and salt marshes that receive increased amounts of nitrogen show increased rates of primary production. The nitrogen budget of *Spartina*-dominated salt marshes on the Atlantic coast of the United States has been investigated in detail, where it has been shown that increased nitrogen supply not only increases the productivity of the plants, but also leads to increased biomass in the detritus feeding invertebrates dependent on the salt marsh (Table 3.4). Over a 2-year period

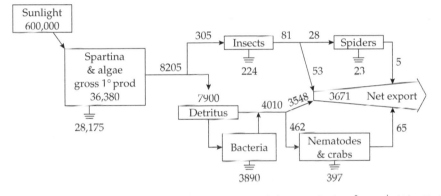

Figure 3.1 Energy-flow diagram for a Georgia salt marsh (units are kcal m^{-2} year^{-1}). (After Teal 1962.)

Table 3.4 Nitrogen budget for Great Sippewissett Marsh. Units are kilogram per year for the entire 0.48 km^2 marsh

Process	Input	Output	Net exchange
Precipitation	380		+380
Groundwater flow	6,120		+6120
Nitrogen fixation (principally bacteria)	3,280		+3280
Tidal water exchange	26,200	31,600	−5350
Denitrification (Total loss of N gas)		6,940	−6940
Sedimentation (0.15 cm year^{-1})		1,295	−1295
Volatilization of ammonia		17	−17
Shellfish harvest		9	−9
Bird faeces	9		+9
Total	35,990	39,860	−3870

Source: Valiela and Teal (1979).

groundwater flow from small underground springs, rainfall, tidal exchange, and the amounts of ammonium, nitrate, nitrite, dissolved organic nitrogen, and particulate nitrogen within each were measured, as well as the fixation of nitrogen by both free-living bacteria, bacteria associated with the roots of marsh plants, and by algae. Measurements were made of the loss of nitrogen from the system due to sedimentation, denitrification, and harvesting of shellfish. Finally, account was made of the input of nitrogen from bird faeces. A remarkably good agreement between the measured input and the output of nitrogen is seen in Table 3.4. The 11% difference is small considering the many possible sources of error in the calculation, and in general it seems that this ecosystem is in balance. Most of the nitrogen budget is controlled by the physical factors of the tide, supplemented especially by groundwater flow. Considerable changes do, however, take place within the salt marsh. For example, 64% of the nitrate, which enters the marsh, is intercepted, and ultimately leaves the marsh in the form of particulate ammonium and nitrogen. An amount of particulate organic matter equivalent to about 40% of the net annual above-ground production of the marsh is exported from this marsh, providing a rich food supply for the detritus feeders. The highly productive salt marsh studied has achieved a balanced steady state, which supports the estuarine ecosystem mainly as a source of particulate organic matter, and as a means of converting and recycling nitrogen.

Relatively little of the *Spartina* is consumed directly by animals, and instead the net primary production of *Spartina* mostly reaches the estuarine ecosystem in the form of fragments broken off the grass. These fragments form the basis for detritus, as bacteria progressively decompose them. By trapping the detritus in the tidal creeks of salt marshes, it has been found that periodic storms are responsible for the export of large quantities of detritus from salt marshes. After one storm over 2000 kg of

detritus was exported in 5 h from a 0.36 km^2 salt marsh. Not all American estuaries receive such large quantities of detritus from *Spartina*. In the salt marshes of the Patuxent river estuary less than 1% of the *Spartina* production reaches the estuary as detritus. This is in great contrast to the 20–45% reported for other estuaries, which is attributed to greater degrees of tidal flooding elsewhere. Dissolved organic carbon (DOC) is also released from the leaves of *S. alterniflora* into the estuarine water, which rhythmically inundates the salt marshes. It has been calculated that the DOC released from *Spartina* is 61 kgC ha^{-1} year^{-1}. Although this represents only a few percent of the total production, the DOC can be readily metabolized by the microbial populations in the water and thus becomes available to consumer animals.

Within British and other north European estuaries the salt marshes are typically found only in the region above the point of the lowest neap high tide. In this region they are not covered by the tide every day, but are covered periodically by spring high tides. Salt marshes thus occur in the upper intertidal area and the plants that occur there must be able to tolerate being covered occasionally by saline estuarine water. Salt marshes display a clear zonation, or successional sequence, from low to high elevations. The plant most typical of the outer, or shore, end of the salt marsh is *Salicornia* (glasswort or marsh samphire). The classical sequence is then *Glyceria maritima*, *Suaeda maritima*, or *Aster tripolium*, above these are *Limonium vulgare* (sea lavender), then *Armeria maritima* (sea pink), followed by *Atriplex* species, and *Festuca rubra* and *Juncus maritimus* toward the top of the salt marsh.

The salt marsh habitat is recognized as a key component of the estuarine ecosystem, and is often specifically protected under legislation. There are, for example, 643 km^2 of salt marsh on European North Sea coasts, over half of which (55%) is on the Wadden Sea coasts, while 26% is on Britain's east coast and 7% is in the Dutch Delta region. The ecosystem role of salt marshes is often identified as a habitat for animals such as juvenile shrimps or fish, and as a roosting and breeding site for birds. For example about 20,000 pairs of Redshank (*Tringa totanus*), being about 60% of the total British breeding population, breed on salt marshes. It is because of their importance to the functioning of the major estuarine systems and their survival of the large populations of winter feeding birds (see Chapter 5) that a very high proportion of salt marshes are protected in Europe. This is not to denigrate their other interests, but their significance as a roost site for birds is a much more obvious manifestation of their conservation value.

3.3 Mangroves

Many of the world's great estuaries are in the tropics. The Amazon, Orinoco, Congo, Zambezi, Niger, Ganges, and Mekong are all very large and receive drainage from enormous catchments. Tropical estuarine environments range in size from tiny seasonally flowing systems of 1–2 km^2 to the estuaries of some of the world's largest rivers. They also encompass extensive coastal lakes and the reduced salinity estuarine waters extending along the coast in parts of southeast Asia, South America, and Africa. In all these tropical and subtropical estuaries mangrove trees occupy a similar habitat to the salt marshes of temperate estuaries, fringing the banks of the estuaries with dense stands of vegetation and forming the dominant intertidal vegetation.

Mangroves generally match the 20 °C isotherms in both hemispheres, suggesting that water temperature is the most significant influence. It has been shown that the presence of mangroves correlates with areas where the water temperature of the warmest month exceeds 24 °C; also that their northern and southern limits correlate reasonably well with the 16 °C isotherm for the air temperature of the coldest month. Latitudinal ranges are greater on eastern continental margins than on western sides due to the presence of warm or cold currents. Tropical estuaries grade into subtropical systems beyond the tropics of Cancer and Capricorn where a winter water temperature low of about 12 °C marks their southern and northern limits. Although many estuaries of the south and south east United States have been described as "tropical" or "subtropical," this is not really the case in a world context, with the possible exception of south Florida mangrove-lined systems. Winter water temperatures in many US Gulf of Mexico estuaries fall as low as 5 °C.

Like salt marshes, the main contribution of mangroves to the estuarine ecosystem is through the abundant supply of plant litter, which is either used directly or in various degraded forms by many animals (Fig. 3.2).

3.4 Intertidal plants and macroalgae

Sea grasses are true flowering plants, and several sea grass species inhabit estuaries, including *Thallasia*, *Posidonia*, and *Cymodocca* in warm and tropical waters, and *Zostera*, *Ruppia*, *Potamogeton*, and *Zannichellia* in temperate areas. Sea grasses, like other estuarine organisms, show greatest species diversity at the seaward and freshwater ends of estuaries, and reduced diversity within the central parts of an estuary. The eel-grass or widgeon grass, *Zostera* spp. is the commonest sea grass on the intertidal estuarine flats in many temperate estuaries growing on sandy and muddy substrata, and occurring subtidally down to 1-m depth (Fig. 3.3). The annual net production is about twice the maximum biomass and ranges from 58 to 330 $gC\ m^{-2}\ year^{-1}$, and exceptionally up to 1500 $gC\ m^{-2}\ year^{-1}$.

In warmer waters, such as in Florida or Puerto Rico, *Thalassia* becomes the main sea grass, with biomasses of 20–8100 g dry wt m^{-2} and productivity values of 100–825 $gC\ m^{-2}\ year^{-1}$. These high productivity values will often be supplemented by 20–30% epiphytic plants, that is smaller plants growing attached to the *Thalassia*. As for other plants, the energy may be utilized by animals, not so much by

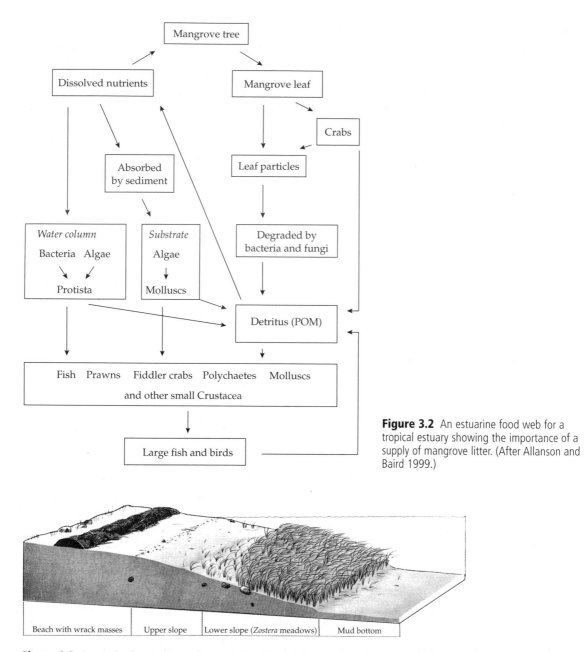

Figure 3.2 An estuarine food web for a tropical estuary showing the importance of a supply of mangrove litter. (After Allanson and Baird 1999.)

Figure 3.3 A typical eel-grass (*Zostera*) community of Danish fjords and land-locked brackish waters, in an area not subject to destruction. (From Rasmussen 1973.)

grazing the sea grass, but rather through the detritus route. The distribution of submerged vascular plants is determined principally by the presence of shallow (sandy) sediments and the turbidity of the water. Shallow, quiet waters, and coastal lagoons with favorable light conditions are the primary sites for the development of submerged sea grasses.

The production of seaweeds (macroalgae) such as *Fucus* and *Ascophyllum* can be high on marine rocky shores. In estuaries, however, populations of seaweeds tend to cover a very small proportion of the total area, being confined to rocky outcrops, quays, and piers. The seaweed *Fucus ceranoides* is confined to estuaries, in contrast to other *Fucus* species that tend to occur only on fully marine coasts. In a comparison of the estuarine *F. ceranoides* with the marine *Fucus vesiculosus* it has been found that the distribution of the species are limited by salinity, with low salinity unfavorable for *F. vesiculosus* and high salinity unfavorable for *F. ceranoides*. The seaweeds of Florida estuaries, which are tolerant to a wide range of temperature, light, and salinity and even short exposures of freshwater, can continue to photosynthesise while both covered and exposed by the tide.

The mudflats of estuaries, which receive high nutrient (especially nitrogen) inputs from inland areas, for example, the Eden, the Ythan, or Chichester harbour in the United Kingdom, may become covered in profuse growths of the green alga *Enteromorpha* (mainly *Enteromorpha prolifera*), which develop as mats during the summer season, and decline in the autumn. At the end of the growing season large populations of heterotrophic bacteria, and subsequently denitrifying bacteria, develop on the rotting algae. The *Enteromorpha* can be the main means of accumulating nitrogen from the waters that flow into the estuary, and as the algal mats decay the nitrogen is made available to other parts of the ecosystem. The mats may also smother the animals living within the mudflats, and as the mats decay they may utilize much of the available oxygen, to the detriment of the animals.

Macroalgae can have high rates of primary production within the areas where they occur (Table 3.5) and on an estuary-wide basis can contribute up to 27% of total primary production. Highest biomass and production values seem to occur when hydrodynamic energy is relatively low, such as in lagoons and tidal inlets. In open river dominated estuaries, especially in funnel shaped estuaries, biomass seems to be lower, although wherever hard substrates occur in the intertidal, locally high densities of macroalgae can be found.

Table 3.5 Net primary production of selected estuarine habitats

	Locality	gC m^{-2} year^{-1}	g dry wt m^{-2} year^{-1}
Zostera	Leaves, Denmark		856
	Below ground, Denmark		241
	Range	58–330	116–680
Thalassia		825	
Salt marsh	*Limonium*, UK		1050
	Salicornia, UK		867
	Baltic meadow		230
	Oregon, USA		1200–1700
	Carex, Oregon, USA		2600
	Netherlands	100–500	200–1000
Salt marsh	Georgia, USA		3700
(*Spartina*)	N. Carolina, USA		650
	United Kingdom		970
	Range	133–1153	500–3500
Mangrove	Range	0–2700	
Comparisons	*Laminaria*	1200–1800	
with the Sea	Coastal plankton	100	
	Open sea plankton	50	

Data from various sources.

3.5 Microphytobenthos

Large populations of diatoms and other *microalgae*, known as *microphytobenthos* or *epibenthic algae*, occur in the upper 1-cm of mudflats, although living diatoms can be found down to 18 cm due to diurnal vertical migration within the sediment. The richest populations of microalgae have generally been found on the lowest parts of the intertidal areas, where the appearance of a "diatom biofilm" can often be very apparent.

In contrast to phytoplankton that typically has pronounced seasonal fluctuations in number and biomass, some authors have found no seasonal fluctuations in the benthic microalgae, due to the continuous regeneration of nutrients by bacteria within the sediment. In the Wadden Sea, Netherlands (Fig. 3.4) and elsewhere, a clear seasonal pattern to the production of the microphytobenthos appears to be closely linked to temperature variations. The microphytobenthos can have a significant role to play in the mudflat estuarine ecosystem, with values of net production of 30–300 gC m^{-2} year^{-1} (Table 3.6). The primary production of epibenthic algae can be compared with the phytoplankton production in the overlying water. The annual net production for the benthic algae in the Lynher estuary, for example, at 143 gC m^{-2} year^{-1} being almost double the value of 81.7 gC m^{-2} year^{-1} for the water column. Much of the epibenthic algae appear to be utilized by bacterial

populations within the mudflat surface and these, together with the algae, are utilized by the consumer animals. Benthic microalgae have a valuable role to play in the formation and maintenance of an oxygenated zone on the surface of intertidal estuarine sediments. Along with the physical forces of the tide, microalgae may be the main source of oxygen for the sediment surface through the process of photosynthesis.

Many questions remain about what factors control microphytobenthic biomass on muddy shores. Suggestions include resuspension, nutrients, grazing, exposure, and desiccation (Underwood and Kromkamp 1999). Indeed, while only a few estimates of the contribution of microphytobenthos production to total estuarine production are available, statements about the importance of microphytobenthic activity in such systems are common. There is little evidence that microphytobenthic assemblages in cohesive sediments are nutrient limited, although carbon dioxide limitation of photosynthesis has been suggested. Microphytobenthic biofilms may play an important role (or barrier) in the exchange of nutrients between the sediments and the overlying water, and thus control bacterial processes within the sediment. Primary production by microphytobenthos is positively related to the elevation of the intertidal flat. The higher the intertidal flat, the longer the emersion period, that is, the longer the photoperiod. Exposure to waves (hydrodynamic

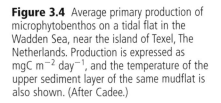

Figure 3.4 Average primary production of microphytobenthos on a tidal flat in the Wadden Sea, near the island of Texel, The Netherlands. Production is expressed as mgC m^{-2} day^{-1}, and the temperature of the upper sediment layer of the same mudflat is also shown. (After Cadee.)

Table 3.6 Microphytobenthic primary production from intertidal sediments in different estuaries

Estuary	Annual production (gC m^{-2})	Mean chlorophyll a (mg m^{-2})	Daily production (mgC m^{-2})
False Bay, Washington	143–226	—	—
Falmouth Bay, Massachusetts	106	—	5–85 (h^{-1})
Bay of Fundy, Canada	47–83	10–500	—
Graveline bay, Mississippi	—	—	5–56 (h^{-1})
North Inlet, South Carolina	56–234	60–120	19–180
Weeks Bay, Alabama	90.1	0.2–30.7	10–750
Ythan, Scotland	116	—	9–226
Wadden Sea, Netherlands	60–140	40–400	15–1120
Lynher, England	143	30–80 µg g^{-1}	5–115 (h^{-1})
Ems-Dollard, Netherlands/Germany	62–276	33–184	600–1370
Oosterschelde, Netherlands (>1985)	242	195	—
Westerschelde, Netherlands	136	113	—
Ria de Arosa, Spain	54	—	—
Tagus, Portugal	47–178	—	5–32 (h^{-1})
Langebaan lagoon, South Africa	63 (sand) 253 (mud)	—	17–69

Sources: Underwood and Kromkamp (1999), Hein *et al.* (1995), and MacIntyre *et al.* (1996).

energy) decreases benthic algal production, and for this reason chlorophyll biomass and hence primary production of microphytobenthos seems to be positively correlated to the clay content of the sediment. That is, fine sediments indicate quiet wave conditions that favor the microphytobenthos.

3.6 Phytoplankton

The phytoplankton is an integral part of the estuarine ecosystem; however, it does not have such a dominant role as, for example, the phytoplankton in marine ecosystems or freshwater lakes. Despite the abundance of nutrients in estuaries, other factors may limit the production of estuarine phytoplankton.

The photosynthesis and respiration of phytoplankton has been measured in a 400-km^2 system of estuaries near Beaufort, North Carolina. The net production was 52.5 gC m^{-2} year^{-1}. While the daily rate of production of phytoplankton could be quite high, the annual rate

is relatively low, which might be due to two factors, shallowness, and turbidity. The shallow nature of the estuaries studied, which is typical of most estuaries, meant that the mean depth of the water, at 1.18 m, was 1.7 m less than the optimum depth for producing maximum net photosynthesis. The penetration of light in estuarine waters is severely limited by the turbidity of the water, due to suspended sediments and POM, which will again limit the production of the phytoplankton. Within the Lower Hudson estuary, USA, dissolved inorganic nutrients are high throughout the year, but large blooms of phytoplankton do not occur despite this availability of nutrients. It is suggested that this is due to the flushing rate of the estuary, whereby the populations of phytoplankton are carried out to sea before their growth rates permit the development of phytoplankton blooms. Those peaks of phytoplankton that do occur are related to the incursion of marine water carrying plankton into the estuary.

Research workers studying different estuaries have come to widely different conclusions regarding the role of phytoplankton, some claiming that primary production of phytoplankton is insignificant, while others regard phytoplankton production as being of central importance to the estuarine ecosystem, responsible for approximately 85% of total ecosystem primary production. Although nutrients appear to be available for the production of large quantities of phytoplankton, maximal production is apparently rarely achieved due to three factors. First, turbidity can limit the penetration of light, second, the shallowness of many estuaries means that blooms may not develop, and third, the growth rate of the phytoplankton may be less than the flushing rate of the estuary. Table 3.7 summarizes the results of measurements of phytoplankton primary production from several estuaries worldwide. The productivity of estuarine phytoplankton may be lower than production values for coastal phytoplankton, and much lower than values for salt marsh production, but the high productivity in relation to the biomass is pronounced, and the phytoplankton is for many animals a richer source of food than plant fragments prior to decomposition. The ciliate protozoa, with their rapid turnover time (2 days) may often be the main consumers of phytoplankton in brackish waters, consuming more than the heavier, but slower-growing, copepods of the zooplankton.

Table 3.7 Phytoplanktonic primary production in different estuaries. Where different areas within estuaries have been studied, values are given from central areas

Estuary	Annual production (gC m^{-2})	Mean chlorophyll a (mg m^{-3})	Max. daily production (gC m^{-2})	Max. chlorophyll a (mg m^{-3})
United States				
San Francisco Bay (South Bay)	150	—	2.2–3.3	10–70
San Francisco Bay (Suisun)	93–98	1.1–14.6	<0.55	20–50
Tomales Bay	360–810	5–11	3–9	15–83
Neuse River	360–531	—	2.5–3.4	40– >50
Chesapeake Bay	337–782	4.0–10.5	—	15
Hudson River	70–220	18–22	1.2	30–55
Bash Harbor Marsh	11	2.6	12	11
Delaware	296	—	—	40
Peconic bay	162–191	2.9–3.2	2.1–4.6	14
Europe				
Oosterschelde	242–406	3.0–6.6	4–8.5	16–36
Westerschelde	184–197	8.5–11.1	1.2–3.7	26–33
Bristol Channel	49	—	—	2–10
Ems-Dollard	26–154	4.6–9.9	1.6–4.5	13–45
Marsdiep (Wadden Sea)	250–390	6–15	5–6.5	38–42
Asia				
Pearl River	20	3–5	—	

Source: Underwood and Kromkamp (1999).

There is a large interplay of variables influencing the rate of phytoplankton photosynthesis (nutrient or light limitation, osmotic stress) and factors influencing biomass such as grazing, washout, resuspension, and deposition. Phytoplankton in estuaries may experience rapid changes in the type of limitation (nutrients, light) and different physical environments (mixing, salinity) and these changes may influence species composition. Interannual variability in primary production can to a large extent be explained by changing watershed conditions and changing land use, as the watershed and rainfall determine the nutrient and sediment input into estuaries from the land. These inputs can both stimulate primary production when the system is nutrient limited or when the light conditions improve, or it can decrease primary production as turbid sediment-laden water can decrease the light availability or flush out the populations. There is a wealth of evidence that, due to increased land use and the associated nutrient load, many estuaries have undergone eutrophication. The effect may to a large extent be dampened out when grazing by suspension feeders is important. In some cases eutrophication can lead to harmful algal blooms in the phytoplankton. Even in this case, total primary production will not necessarily change, but the changes in nutrient

concentrations and ratios may influence species composition of phytoplankton, which might have profound ecological implications.

Primary production generally increases toward the mouth of an estuary, indicating that the decreases in nutrients are more than compensated for by the increased water transparency. This fact alone demonstrates that primary production is not determined solely by nutrient input and availability. Table 3.8 shows some recently published values of annual production in different regions of several estuaries, and Table 3.9 shows phytoplankton biomass. These tables show a trend for increased annual production toward the outer part of the estuary, but that maximal biomass may occur in inner areas. Light and nutrient availability mainly influence phytoplankton growth. Freshwater input can have negative effects by cresting osmotic stress, by flushing out populations and by increasing turbidity, but can have positive effects as a major source of nutrients and by creating a stratified water column, which can improve light conditions and thus initiate phytoplankton blooms. In view of these conflicting forces, single environmental variables are poor predictors of phytoplankton primary production in estuaries. Primary productivity of estuarine phytoplankton is mainly controlled by three variables: Phytoplankton

Table 3.8 Estimates of annual production of phytoplankton (gC m^{-2}) in several estuaries, arranged as freshwater, inner, central, or outer

Estuary	Tidal freshwater	Inner	Central	Outer
Fourleague bay, USA	322	514		
Tomales bay, USA		70	420	460
Hudson River, USA	70–240			200
Delaware Estuary, USA		105	296	344
Peconic bay, USA		213	177	
Bristol channel, UK		7	49	165
Ems Dollard, Germany, Netherlands		70	91	283
Westerschelde, Netherlands	388	122	184–197	212–230
Oosterschelde, Netherlands		301	312	382

Source: Heip *et al.* (1995) and references therein.

Table 3.9 Estimates of maximum biomass of phytoplankton (mg chlorophyll m^{-3}) in several estuaries, arranged as freshwater, inner, central, or outer

Estuary	Tidal freshwater	Inner	Central	Outer
Chesapeake bay, USA		<25	<25	<18
Columbia River estuary, USA		<15	<7	<5
Great Ouse, England		<70	<50	<10
Bristol channel, UK		<3	<9.2	<1.6
Shannon, Ireland		<70		<5
Westerschelde, Netherlands	55	10–20	8	8
Oosterschelde, Netherlands		<30–50	<30	<40

Source: Heip *et al.* (1995) and references therein.

biomass (i.e. chlorophyll concentration), incident irradiance, and turbidity according to Heip *et al.* (1995). Nutrient concentration, grazing, transport, sedimentation, temperature, and daylengtth seem to be of minor importance. Others caution against over-emphasizing the role of turbidity and under-estimating the effects of nutrients and suggest that while turbidity probably controls productivity in macrotidal systems or in river-dominated reaches, nearly all estuaries experience some degree of nutrient limitation near their seaward boundaries.

3.7 Detritus

Detritus has already been defined as all types of biogenic material in various stages of microbial decomposition. Much of this biogenic material may be fragments of plants, such as *Spartina*. In estuaries without large salt marshes the main sources of detritus are fragments of dead plants and animals from the sea, from rivers, or from the estuary itself, as well as the faeces, and other remains of the estuarine animals. All the types of primary production described in the preceding sections of this chapter can supply material that becomes detritus, and it is clear from many studies that most primary production in estuaries is not consumed directly by herbivores, but rather is converted into detritus before consumption by detritivores.

In the context of an estuarine energy budget, detritus is referred to as allochthonous sources of particulate organic carbon, to distinguish it from autochthonous material, being that produced by the primary producers through photosynthesis. Allochthonous sources can be divided into (a) riverine, (b) marine, (c) atmospheric, and (d) erosion inputs, as well as (e) direct domestic and industrial inputs. The relative importance of the various sources depends on factors such as river discharge, tidal amplitude, estuarine morphology, land usage, and human population as well as the geology of the area. Inevitably the proportions of the different sources will vary from estuary to estuary. In the context of allochthonous organic matter it is important not only to recognize the source and amount of organic matter but also its "quality." *Labile* allochthonous organic matter refers to material, which can be readily degraded and hence made available to consumers, whereas *refractory* matter is obstinate and resistant to degradation and hence may be unavailable to consumers.

Spartina and other plant detritus is relatively indigestible to the consumer animals and thus much of the flux of organic matter to detritivores must involve the conversion of the particulate detritus to soluble compounds and their assimilation by microorganisms, which can then be consumed by detritivores. The nutritive value of *Spartina* increases as the detrital fragments become enriched with microbial populations. In Fig. 3.5 it can be seen that living *Spartina* has a content of 10% protein. Dead leaves entering the water have about 6% protein, but as the plant fragments

become smaller the protein content increases to 24%. Thus the detritus, which is rich in protein, may be a better food source for animals than the grass tissue that formed the basis for the particulate matter. Similar results have been described for leaves of the tropical estuarine salt-marsh plant, red mangrove, with 6.1% protein in leaves on the tree, 3.1% protein at leaf fall, and 22% protein after decomposition in estuarine water for 12 months. The main decomposers of plant material in seas and estuaries are bacteria, as shown in Fig. 3.6.

The detritus, composed of the decaying remains of plant primary production, and microbes, has a valuable role in stabilizing the estuarine ecosystem by leveling out the seasonal variations in primary production, ensuring a year-round food supply, and securing the reabsorption of dissolved nutrients. The role of microorganisms in the process of the breakdown of plant material in estuaries may be compared to the role of microorganisms in the guts of terrestrial herbivores. The bacteria living on particulate or dissolved organic matter in both cases make the primary production more readily available for animal consumption.

When bottom-dwelling animals consume detritus, it appears that they consume the bacteria and other microbes, but reject the plant tissues. In the process they may shred the plant material into finer fragments, which will

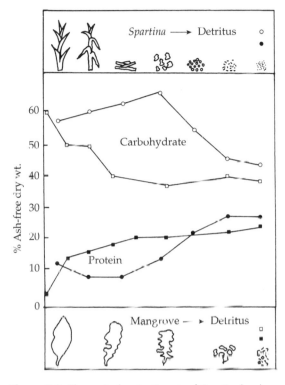

Figure 3.5 Change in the constituents of *Spartina* (o, ●) and Red Mangrove (□■,) leaves during conversion from living plant material to fine detritus fragments, as shown pictorially. (After Odum and de la Cruz 1967; Heald 1969.)

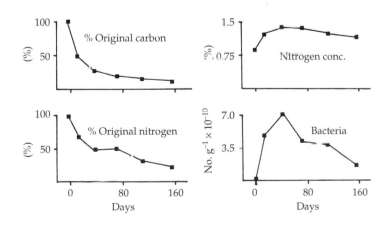

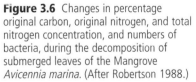

Figure 3.6 Changes in percentage original carbon, original nitrogen, and total nitrogen concentration, and numbers of bacteria, during the decomposition of submerged leaves of the Mangrove *Avicennia marina*. (After Robertson 1988.)

provide a larger surface area for microorganisms, and so accelerate the processes of decay. The activities of animals that consume the entire sediment, lead to a continual mixing of the organic and inorganic particles in the sediment, in a process known as bioturbation. This will tend to distribute detrital material throughout the surface layers of the sediment, and so enable material, which has settled on the surface of the sediment to organically enrich the sediment to a depth of several centimeters. It has been calculated that the biomass of bacteria within estuarine sediments may be of the same order of magnitude as the biomass of the animals in the sediment. Apart from the immediate surface layer, estuarine sediments tend to be anaerobic, as the bacteria and other microorganisms consume all the available oxygen. Much detritus therefore undergoes anaerobic metabolism, with hydrogen sulfide, methane, or ammonia produced, as well as dissolved organic carbon compounds that can be utilized by aerobic microorganisms living on the surface. These aerobic microorganisms may also be consumed by detritivores.

It should not be assumed that all detritus is the same for detritivore animals. Studies on the utilization of detritus by the polychaete *Capitella capitata* have shown that detritus derived from salt marsh and sea grass plants that contain a high percentage of unavailable energy is consumed as the products of microbial decomposition and protein enrichment, whereas detritus derived from seaweeds may be consumed directly.

The primary consumers that ingest POM from the water column do so without regard to whether it is phytoplankton, suspended detritus, or microbial organisms. One study attempted to segregate these components, and came to the conclusion that the bay scallop, *Argopecten irradians*, ingests 20% phytoplankton and 80% detritus and bacteria, emphasizing the much greater availability of detritus within the estuarine ecosystem, even though the growth rate of the scallops would have been higher on a diet of phytoplankton alone. The

activities of these suspension-feeding bivalves may be profound. In South San Francisco Bay estuary, for example, the suspension-feeding bivalves filter a volume equivalent to the total volume of the area each day, and this grazing may be the primary mechanism controlling phytoplankton biomass. In Chesapeake Bay, USA, the detritus averages 77% of the total organic particles in the water column, and the phytoplankton 23%. The numbers of detritus particles in the water showed little seasonal variation, whereas the phytoplankton in this area showed considerable seasonal variation.

Estimates of the global organic inputs into estuaries indicate that the major sources are primary production from both wetlands (salt marshes) and planktonic and intertidal algae, along with organic matter carried into the estuary from rivers. To these values must be added man's discharges of sewage, oil products, food products, and wood pulp and a quantity of organic matter entering the estuary from the sea. All these sources of organic matter are utilized by microorganisms within the estuaries of the world, to become detritus.

Along with the POM that forms detritus in estuaries, there may be considerable quantities of dissolved organic matter present, derived from plant exudation, animal excretion, and from the products of decomposition. The capacity for the uptake of dissolved organic matter by animals is widespread, but despite this it seems likely that estuarine animals get the vast majority of their food from POM. The dissolved organic matter will mainly be metabolized by bacteria, and some estimates show that bacterial production utilizing dissolved amino acids can amount to 10% of algal production. The bacteria, consuming the dissolved organic matter, themselves become part of the particulate matter in the estuary.

In parallel with the decomposition of organic matter there is liberation of carbon dioxide, phosphorus, nitrogen, and other nutrients. This recycling of nutrients, referred to as mineralization, is a prerequisite for the new production of organic matter by autotrophs.

3.8 Summation of plant and microbial production in estuaries

The various components of primary and microbial production can be combined in an attempt to understand a particular estuary, and to try and explain the high productivity of estuaries in general.

Table 3.10 summarizes the data on primary production from 12 estuaries, where the various components have been measured, and allows us to examine the relative contributions of the various producers. It must first be emphasized how variable the total production is, with total production ranging from 63.6 to 1600 gC m^{-2} year^{-1} and the examples given are from various latitudes with conditions ranging from mangrove and *Spartina*-dominated

estuaries, through to estuaries dominated by bare mudflats. Phytoplankton production contributed between 2.2 and 43.3%, while epiphytes were less than 8.5% where studied, and macroalgae (mainly fucoids) also contributed little, except in Flax Pond, where they supplied 20.5% of net production. In the estuaries with bare mudflats, epibenthic algae contributed over 30% of production, but in those dominated by *Spartina* it was much less. When present, *Spartina* supplied up to 84% of total primary production. From these various studies, it must be concluded that each estuarine ecosystem has its own characteristics, with a unique mix of primary producers. As far as the primary consumers are concerned, the mix of primary producers may not be very important, if most energy is consumed in the form

Table 3.10 Net primary production of particulate material in various estuaries, expressed as percentage of total production (gC m^{-2} year^{-1})

Estuary	Phyto-Plankton	Microphyto-Benthos	Zostera	Epiphytes	Spartina	Salt marsh	Mangrove	Total Prod.
Beaufort, North Carolina	43.3	NR	38.0	8.5	9.8	0	0	152.6
Flax Pond, New York	2.2	5.6	0	3.7	74.5	20.5	0	535.0
Sapelo Island, Georgia	6	10	0	0	84	0	0	1445
Nanaimo, Canada	11.8	39.5	42.1	0	0	6.6	0	63.6
Barataria bay, Louisiana	18.9	22.2	0	0	69.0	1.1	0	890
Waitema Harbour, New Zealand	25.3	30.8	0	0	0	0.3	31.4	473.5
Wadden Sea, The Netherlands	50	50	0	0	0	0	0	—
Lynher, England	33	33	0	33 (Detritus)	0	0	0	—
Langebaan Lagoon, South Africa	23	22	55 (macro-phytes)	0	0	0	0	—
North Inlet, South Carolina	12	29	0	13 (macro-algae)	46	0	0	1600

Source: Knox (1986) and Underwood and Kromkamp (1999).

of detritus, and it may be the supply of detritus derived from the breakdown of the primary producers, which is the feature of most importance to the success of the primary consumers.

Measurements have been made in the Dollard estuary on the Dutch–German border in an attempt to quantify all sources of organic input to this estuary. Eighty percent of the estuary is composed of tidal sand/mudflats. The total amounts of organic carbon entering and leaving the Dollard estuary are shown in Table 3.11. This preliminary attempt at a carbon budget has clearly revealed a large discrepancy between the measured inputs and outputs, which is believed to be mainly due to unquantified export of dissolved carbon from the estuary. Nevertheless, several valuable points emerge from this study. First, the main sources (75%) of carbon are outside the estuary in the river, the sea, and an industrial plant (potato flour mill), which discharges effluent. Second, due to the turbidity of the water the primary production from phytoplankton is only 7.5% of the primary production from benthic algae such as diatoms and blue-green algae. Furthermore, the total primary production of 10×10^6 kgC year^{-1} is considerably less than the carbon consumed, or utilized

Table 3.11 Organic carbon budget for the Dollard estuary (units are $\times 10^6$ kgC year^{-1} for the entire area of almost 100 km^2)

Import/ production		Export/utilization of organic carbon	
Particulate C from North Sea + River Ems	37.1	Dissolved C to North Sea	7
From potato flour mill	33.0	Utilization in water	7.2
From salt marshes	0.5	Utilization in sediment	18.2
Primary production Phytoplankton	0.7	Buried in sediment	9.9
Primary production Microphytobenthos	9.3	Bird feeding	0.26
Total import	80.6	Total export	35.56

Source: van Es (1977).

in the water and sediments (25.4×10^6 kgC year^{-1}).

The study of the Dollard estuary clearly shows that primary production within an estuary is inadequate to support the large number of detritus feeders inhabiting the mudflats, and the detritus feeders must rely on the importation of organic debris from outside the estuary. What primary production does take place is due overwhelmingly to the benthic algae, rather than the phytoplankton whose production is inhibited by the turbidity of the water. Thus the basic biological processes creating energy for the primary consumers in this estuary are concentrated on the mud surface with the primary production of the benthic algae, and the transformation of organic debris into more digestible material by bacteria. The high productivity of the Dollard, and many similar estuaries, is thus seen to be due to the position of the estuary as a collecting area for organic matter, supplemented by the primary production of the benthic algae on the intertidal flats. The high productivity of this estuarine ecosystem is thus because it is subsidized by the transfer of energy from other ecosystems.

The Grevelingen estuary, in the Netherlands, was studied intensively prior to the implementation of the Delta Barrage scheme, which is described in Chapter 6. This estuary covered 140 km^2, of which 81 km^2 was covered at all times, 55 km^2 was intertidal sand and mudflats, and 4 km^2 was salt marsh. A detailed food budget for the period before it was dammed is shown in Table 3.12. The levels of primary production are dominated by the production of phytoplankton, supplemented by benthic microalgae. This is the reverse of the situation in the Dollard estuary, and is due to the less turbid waters in the Grevelingen, coupled with the smaller proportion of intertidal area. The total contributions from all sources within the estuary to the carbon budget is, however, exceeded by the material carried in on each tide from the adjacent North Sea.

The carbon budget for Barataria Bay, Louisiana, USA shown in Table 3.13 reveals

Table 3.12 Food sources for the Grevelingen estuary, The Netherlands, expressed as the import and production of particulate organic carbon

Food source	gC m^{-2} year^{-1}
From salt marshes	0.3–7
Production Zostera	5–30
Production microbenthic algae	25–37
Production phytoplankton	130
Land run-off	2
Import from North Sea	155–255
Total of sources	317–451

Source: Wolff (1977).

Table 3.13 Carbon budget for Barataria Bay, Louisiana, (gCm^{-2} year^{-1}) assuming that 1 gC is equal to 2 g dry organic wt

Source		Sink	
From saltmarshes	297	Consumed in estuary	132
Production benthic algae	244	Exported to sea	318
Production phytoplankton	209		
Totals	750		750

Source: Wolff (1977).

Figure 3.7 The Forth estuary, eastern Scotland, UK. A typical "European"-type estuary with large intertidal mudflat areas, bare of macrophyte vegetation. Microphytobenthos is a main primary producer in such habitats, in the foreground, the plant *Salicornia* can be seen colonizing the uppermost areas of the mudflat.

that unlike the previous European examples, it is a net exporter of energy rather than a net importer. While the primary production within the estuary of phytoplankton and benthic algae is important, the largest source of energy is the supply of detritus from the *Spartina*-dominated salt marshes.

From a consideration of the energy budgets presented above, it is clear that two distinct types of estuary emerge, although there is undoubtedly a spectrum of types, with the most distinct examples at the opposite ends of the spectrum. At one extreme are the "European-type" estuaries, such as the Dollard, which are dominated by large, relatively bare intertidal mudflats (Fig. 3.7), and at the other extreme are "American-type" estuaries, which are dominated by large stands of the marsh grass *Spartina* (Fig. 3.8).

Figure 3.8 The Great Bay estuary, New Hampshire, USA. A typical "American"-type estuary where the macrophyte *Spartina* occupies much of the intertidal habitat. In the foreground, fragments of *Spartina* are decomposing, and ultimately supplying detritus for the ecosystem.

In the "European-type" estuary much of the primary production within the estuary is performed by large populations of microscopic benthic algae living on the surface of the mud supported by phytoplankton in the water column. The extent of the primary production of the phytoplankton depends on the turbidity of the water. However, in these estuaries the majority of the energy within the primary producer trophic level is derived from outside the confines of the estuary, and is in the form of organic matter, which is carried into the estuary, usually from the sea, but also from land discharges of river water or sewage. The estuary is thus a net recipient of energy, and the high productivity that supports large populations of consumer animals is due to the position of the estuaries as traps for both nutrients and POM.

In the "American-type" estuary the primary production of benthic algae and phytoplankton is important for the productivity of the whole ecosystem, but the dominating factor is the much greater proportion of the estuary, which is inhabited by rich beds of *Spartina* grass. The *Spartina* is only consumed directly by animals to a small extent, and instead they rely on the fragments of *Spartina* forming the substrate for large populations of bacteria, which form detritus, which is then ingested by the animals. Despite high rates of consumption within the estuary, excess material remains, which is carried out of the estuary to fertilize the adjacent sea.

For both types of estuary, and those intermediate between the two extremes, we can conclude that the high levels of production within estuaries are due to a plentiful supply of nutrients supporting the primary production of benthic algae, phytoplankton, and salt marshes. This production is enhanced by the import of POM into the estuary from either the sea or the margins of the estuary, which undergoes microbial decomposition within the estuary to yield a rich food supply for the consumer animals.

In all estuaries the gradients in concentrations of nutrients and turbidity are steep and ecosystem studies have often emphasized the light limitations on pelagic systems imposed by estuarine turbidity and that benthic primary production can therefore be relatively very important. Nevertheless, the estuaries are large net exporters of excess nutrients. Estuaries are also large net importers of carbon (mainly as detritus). The primary transfers of energy within estuarine ecosystems therefore derive from organic detritus inputs plus microphytobenthos to support benthic communities that in turn support the birds, fish, and shrimps, as will be seen in the next chapters.

CHAPTER 4

Primary consumers: herbivores and detritivores

4.1 Introduction

Estuaries are rich in food sources for the primary consumer trophic level in the food web. The main food source is the large quantity of detritus, which abounds in the water column and on the bottom of the estuary. This supply of food is replenished by both tides and river inflow, and the deposition of fine particulate matter and detritus in the central reaches of the estuary provides a food store that is available for virtually the whole year. Whereas food chains in temperate seas and freshwater lakes are dominated by short bursts of primary production, especially in the spring, estuaries are characterized by having food sources available for the whole year. The food sources are, however, richer in the spring and summer as increased temperatures accelerate all biological production.

Most of the primary consumer animals are found on the bottom of the estuary, where a rich benthic community usually develops. Zooplankton is present in the water column, but the strong tidal currents and river flow, which flush out the estuary, coupled with the limitations imposed on phytoplankton by turbidity, make zooplankton a less dominant feature of the estuarine food web than food webs in the sea. For both planktonic and benthic animals, most food is in fine particulate form, whether it is live phytoplankton or the variously decomposing fragments, which make up the detritus. The bacteria-rich detritus is regarded as a better source of food for the primary consumer than the plant tissues that formed the original material for much of

the particulate organic matter (POM). It is generally considered that the major food source for deposit-feeders are the microbes attached to sediments and detritus particles, but it has been difficult to specify the actual food sources for individual animals. The food ingested can either be derived from "microbial stripping," or from amorphous organic debris. It seems likely that both sources are required to support deposit-feeders. The food available cannot be equated with organic carbon, as the supply of nitrogen, amino acids, fatty acids, and vitamins may be just as important. Various studies have concluded that although bacteria can be very important, alone they are insufficient to support the macrobenthic deposit-feeders, and that nonliving detrital material should be regarded as the prime energy source for subsurface deposit-feeders. The detrital material thus forms the main carbon source for these animals, but the bacteria and other microorganisms living in the sediment provide essential protein nitrogen, fatty acids, and vitamins.

It is possible to attempt to classify the benthos into suspension (or filter) feeders relying on fine particulate matter suspended in the water, and deposit-feeders that rely on the food contained within the muddy estuarine deposits. The borderlines between these two groups are, however, often unclear in the estuarine ecosystem. Among the suspension-feeding members of the estuarine benthos, the most typical are the bivalve molluscs, such as the common mussel, *Mytilus edulis*, which rely on phytoplankton and small organic particles. Their filtering process accepts food particles

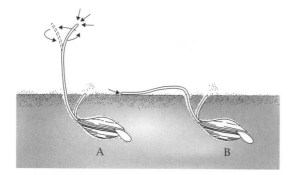

Figure 4.1 *M. balthica*—The left individual is seen feeding on suspended matter with the inhalent siphon stretched up in the water. The right individual is seen sucking in bottom material with the inhalent siphon. For both, the exhalent siphon rejects waste material. (From Olsen, in Rasmussen 1973.)

Table 4.1 Typical densities of estuarine macrobenthic fauna (all as number m^{-2}) in two European estuaries

Species	Feeding mode	Oosterschelde, Netherlands	Forth estuary, Scotland
Mollusca			
C. edule	Suspension	260–5565	200–300
M. balthica	Suspension and Deposit	120–727	78–500 (Spat 2000)
S. plana	Mainly suspension	12–13	0
Mya arenaria	Suspension	2–17	30–40
H. ulvae	Grazing and deposit	1060–14,000	400–153,000
Annelida			
L. conchilega	Suspension	35–71	
Nephtys sp.	Carnivore	58–77	<160
Hediste (= Nereis) sp.	Carnivore and Suspension	235–425	36–750
Crustacea			
C. volutator	Deposit	173–368	15–1985

Compiled from various sources.

within a specific size range (e.g. >2 μm for *Mytilus*) and it is unable to distinguish between phytoplankton and floating detritus of similar size. Among deposit-feeders, the typical example may be the lugworm, *Arenicola marina*, which ingests large quantities of mud, and having digested off any organic material, rejects the bulk of the mud in the familiar worm cast. Between these clear-cut examples are many animals that can both suspension-feed and deposit-feed. The Baltic tellin, *Macoma balthica* lies buried in the mud with its siphons protruding (Fig. 4.1). When food is available in the water overlying the mud, it is able to raise the incurrent siphon and draw in water and particulate matter. When detritus particles are available on the surface of the mud surrounding its buried location, it can reach out across the mud and collect the particles by using its siphon as a "vacuum cleaner." Within the animal the same gills provide a size-limited filtration for the food particles, so that *Macoma*, for example, can spend 10–40% of its life suspension feeding and 60–90% deposit-feeding.

The numbers (Table 4.1) of suspension-feeding benthic animals appear to fluctuate more, than those of deposit-feeders do. This is because the food supply for suspension-feeders, which is phytoplankton and suspended POM, varies seasonally during the year, or even from day to day. Whereas the food supply for deposit-feeders, which is benthic algae or detritus and the microbes growing on it, will vary less throughout the year. Furthermore the biomass, but not necessarily the abundance of suspension-feeders shows a gradation within the estuary, with a greater biomass of suspension-feeders near the seaward end of the estuary, in order to feed on material brought in by the tides (Fig. 4.2).

As an alternative to the classification of the benthos into suspension-feeders and deposit-feeders, the system of epifauna (or surface fauna) and infauna (or buried fauna) may be suggested. Typical epifauna are animals such as mussels (*Mytilus*) on a mussel bed, barnacles

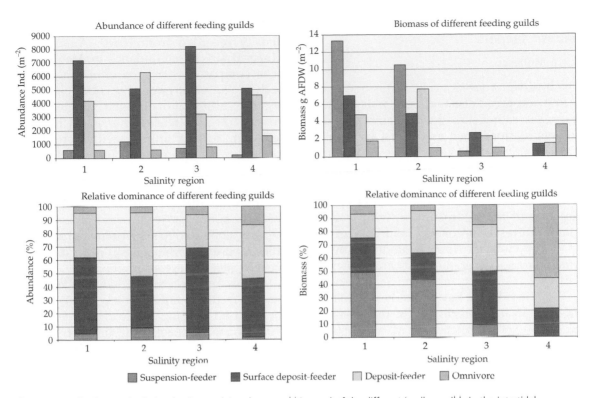

Figure 4.2 Absolute and relative dominance (abundance and biomass) of the different feeding guilds in the intertidal zone of each salinity region of the Oosterschelde estuary. Region 1 is nearest the mouth, and Region 4 is nearest the head of the estuary. (After Ysebaert *et al.*, 2003.)

(*Balanus*), and winkles (*Littorina*) on rocky outcrops, or mobile animals such as mysid shrimps (*Neomysis*) and gammarid amphipods (*Gammarus*). Among the infauna are clearly many of the estuarine worms, such as the polychaetes *Nereis*, *Nephthys*, *Arenicola* or the oligochaetes, *Tubificoides*, *Tubifex*. Other common animals spend some of their time buried in the mud, and then emerge onto the surface to feed, such as the small gastropod snail *Hydrobia*, and the crustacean amphipod *Corophium*.

When examining estuarine benthic samples it is frequently desirable to sort them on the basis of size. Animals retained by a 500 μm (0.5 mm) sieve are classified as macrobenthos, animals retained by a 40–60 μm sieve classified

as meiobenthos, and animals passing a 40–60 μm sieve classified as microbenthos.

In order to examine the productivity of this important primary consumer trophic level we shall examine first the macrofaunal mud dwellers, which are mostly infaunal, deposit-feeders, but which may make excursions into the epifauna or utilize the suspension mode of feeding. Then we shall consider the macrofaunal surface dwellers considering both mobile and static representatives who are predominantly suspension-feeders. The potentially important, but less studied, meiobenthic fraction will also be considered, as will the zooplankton. Finally, the reasons for the high levels of production will be examined.

4.2 The mud dwellers—benthic deposit feeders

4.2.1 Molluscs

The bivalve *M. balthica* (Fig. 4.1) is one of the most widespread of estuarine benthic animals. It can be a rather slow-growing animal, and in some situations such as the St Lawrence in Canada may take 12 years to grow to 14 mm. In Massachusetts or the Wadden Sea it grows faster, reaching 26 mm in 6 years. It is proposed that growth rate and longevity are a function of the hydroclimate; with warmer temperatures a larger size and shorter life span are observed. Typical productivities of Macoma and other molluscs are shown in Table 4.2

The Minas Basin, Bay of Fundy, Canada has tides of up to 17-m range. On the intertidal mudflats exposed by the tide, *M. balthica* is the dominant organism, along with the amphipod *Corophium volutator*, and adult *Macoma* can be as abundant as 3500 m^{-2}, which is the highest density recorded from North America. At Minas Basin, the abundance of *Macoma* has been compared with environmental factors such as sediment grain size, tidal elevation, organic carbon and nitrogen, and bacterial density. The sediment grain size was found to set limits for the animal, with none being found in sands coarser than 0.23 mm, or finer than 0.024 mm. Within these limits it was found that the density of *Macoma* could be accurately predicted by bacterial density alone. *Macoma* cannot survive by deposit-feeding on bacteria alone, and to survive in the large densities reported at Minas Basin it must supplement its diet by suspension-feeding.

Macoma balthica in the Wadden Sea feeds as a deposit-feeding animal for up to 90% of the time that it is feeding. While deposit-feeding it is utilizing mainly benthic algae, which grow on mudflats. *Macoma* growth starts in the spring as soon as the production of microphytobenthos

Table 4.2 Productivity (gC m^{-2} year^{-1}) of estuarine molluscs

Species	Productivity (gC m^{-2} year^{-1})	Location	Habitat
Bivalves			
M. balthica	5.03	Ythan, Scotland	Mudflat
	1.95	Bay of Fundy, Canada	Mudflat
M. edulis	134	Ythan, Scotland	At mussel bed
	24.3	Ythan, Scotland	Whole estuary
	1875	Morecambe bay, England	At Mussel bed
	87.3	Dublin Bay, Ireland	At Mussel bed
	55.5	Wadden Sea, Netherlands	Mussel bed
M. arenaria	5.8	Petpeswick, Canada	Mudflat
Modiolus demissus	5.75	Georgia, USA	Salt marsh
C. edule	40.3	Wash, England	Sand flat (in patches)
	31.0	Baie de Somme, France	Muddy sands
Crassostrea virginica	171.2	South Carolina, USA	Oyster reef
S. plana	12.4	Bangor, Wales	Mudflat
Gastropods			
Hydrobia ulvae	6.4	Grevelingen, Netherlands	Mudflat
Littorina saxatilis	0.75	Petpeswick Inlet, Canada	Spartina marsh

Source: Wilson (2002) and McLusky (1989).

reaches 100 mgC m^{-2} day^{-1} and continues throughout the spring and early summer as primary production increases. The annual growing season starts earliest in years that follow a mild winter. As the temperature reaches 10 °C spawning occurs, and growth ceases at temperature above 16 °C. Thus growth is restricted to a range of water temperatures between 4 and 16 °C in spring. The variation in the *Macoma* population in the Wadden Sea over an 8-year period has been shown to be related to the level of the food supply coupled with the time of immersion by water. Other studies agree that *Macoma* behaves most of the time as a deposit-feeder, but have shown that it depends for its food intake for the greater part on food present in the water column. The explanation for this apparent contradiction is that, as shown in Fig. 4.3, much of the total food intake is rejected as being unsuitable, and is lost as feces (or pseudofeces). A greater proportion of the food taken in by suspension-feeding, rather than deposit-feeding, is regarded as suitable and hence it is this food, which forms the main basis of their diet.

Hydrobia ulvae (Fig. 4.4) is an important feeder on detritus and algae on the surface of mudflats, although it retreats into the surface layers of the mud when the tide recedes. *Hydrobia* is a selective deposit-feeder that feeds mainly by grazing on diatoms growing on the surface of particles of 20–250 μm diameters. Alternatively it can secrete mucus, which traps bacteria, and then it reingests the enriched mucus. Eutrophication by sewage effluent has led to the development of extensive algal mats over of the former open mudflats in several estuaries. The presence of the mat reduces the biomass and diversity of the infauna, but leads to a great increase in the numbers and biomass of *Hydrobia*. In one study at Langstone harbor the biomass of *Hydrobia* increased from 5.4 g m^{-2} to 27.4 g m^{-2} with abundances increasing from 9000 m^{-2} to 42,000 m^{-2} due to the algal mats, which provide a rich food supply for the *Hydrobia*. The hatching of embryonic snails of *H. ulvae* can occur at salinities from 8 to 60 and at temperatures up to 35 °C. The newly hatched snails attach to the surface of the adults shells. The wide range of possible salinity and temperature combinations for hatching coupled with the non-planktonic development and the variable modes of feeding of the adults are all important reasons for the success of this species in estuaries.

The survival strategies for estuarine molluscs are, thus, a tolerance of low salinities, a tendency toward non-pelagic larvae, and the

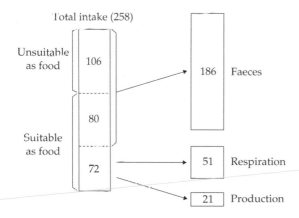

Total intake (258)

Unsuitable as food — 106

186 Faeces

80

Suitable as food

72

51 Respiration

21 Production

Figure 4.3 Schematic model of the main pathways in the energy budget of a population of 1+ year old *M. balthica* on a tidal flat in the Dutch Wadden Sea. All values are in kilo joules per square meter per year. Note that of the total food intake of 258 kJ, 106 kJ were unsuitable as food, and with a further 80 kJ of food were rejected as feces. Only 72 kJ were assimilated, for use in respiration and production. (After Hummel 1985.)

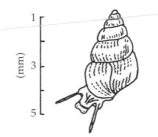

Figure 4.4 *H. ulvae*—This small snail is one of the commonest inhabitants of the surface muds of European estuaries.

adoption of a flexible mode of feeding designed to cope with the problems of a muddy environment. Many estuarine molluscs can utilize a number of different food sources, either at different times in their life history, or at different phases of the tide. This flexibility of feeding strategies is the hallmark of a successful estuarine mollusc.

4.2.2 Annelids

The polychaete annelid worms, commonly known as bristle worms, are the most diverse group of worms in an estuary. Four Nereid polychaete species (*Nereis* (=*Hediste*) *diversicolor*, *Nereis* (=*Neanthes*) *virens*, *Nereis* (=*Neanthes*) *succinea*, *Nereis pelagica*) commonly penetrate into estuaries, and show differences in their physiological tolerances, and in the pattern of their life cycles. The ragworm (*N. diversicolor*) is able to tolerate the lowest salinities, appears to have only a brief (<6 h) larval planktonic phase to assist in retaining the populations within estuaries, and is usually the commonest worm in European estuaries. In America, the clamworm (*N. succinea*) is the most likely species of polychaete to be encountered. Typical productivities of Nereis and other annelids are shown in Table 4.3 from which it is clear that the polychaetes with their shorter life spans and faster growth than, for example, most molluscs are a key component in the productivity of estuaries.

The biomass of the lugworm, *A. marina* (Fig. 4.5), is apparently controlled by the amount of organic matter in the sediment, although it is not found in sediments finer than a median particle diameter of 80 μm, as it cannot maintain its burrow in such fine sediments. In a closely related species *Abarenicola pacfica* it has been found that a "gardening" strategy is adopted whereby the undigested sediment, which is passed out as feces is populated by microbial organisms, which increase the nitrogen content of the sediment by their activity. The lugworm then reingests the enriched feces.

Table 4.3 Productivity (gC m^{-2} year^{-1}) of estuarine annelids

Species	Productivity (gC m^{-2} year^{-1})	Location	Habitat
N. (=*Hediste*) *diversicolor*	27.2	Belgium	Brackish pool
N. diversicolor (=*Hediste*)	5.45	Ythan estuary, Scotland	Mudflat
A. marina	3.13	Grevelingen, Netherlands	Fine sand flat
Nephtys hombergii	3.68	Lynher estuary, England	Mudflat
Ampharete acutifrons	1.16	Lynher estuary, England	Mudflat

Source: Wilson (2002) and McLusky (1989).

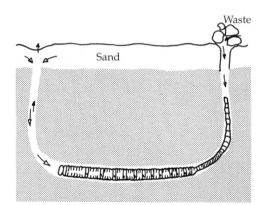

Figure 4.5 *A. marina*—The position of the lugworm *Arenicola* in its burrow. Filled arrows (from waste end) indicate water currents, and open arrows (at head end) indicate the ingestion of sand. (After Kruger 1971.)

The sand-mason (*Lanice conchilega*) occurs on the sandier parts of many estuaries, where it forms burrows up to 30 cm deep but with a small protrusion of about 2 cm standing clear of the sand surface. For much of the time it is a surface deposit-feeder, but it can spend up to 35% of its time feeding on particles in suspension in the water. This combination of two modes of feeding enables it to utilize alternative food supplies and can lead to large populations developing. In the Weser estuary, northern Germany, populations of the polychaete *Lanice* occur at 20,000 animals m^{-2} along with 700 m^{-2} of the suspension-feeding anemone *Sagartia troglodytes*. Their enormous biomass of over 1.1 kg dry wt m^{-2} is only possible because of the strong currents in the area (up to 1 m s^{-1}) carrying in diatoms and plankton for the estuarine benthos to feed upon.

The smaller oligochaete worms have been studied much less than polychaete worms, partly due to their small size, which makes them less conspicuous. However, they are present in large numbers in most estuaries, thriving in the low-oxygen conditions, which typify life on a mudflat. They may become the sole inhabitants of estuarine mudflats under conditions of organic enrichment. A study of the Forth estuary has recorded biomasses of 27.97 g dry wt m^{-2} of oligochaetes from organically enriched areas, and 6.30 g dry wt m^{-2} from more typical estuarine mudflats. The latter populations, dominated by *Tubificoides benedeni*, had an annual production of 25 g dry wt m^{-2} year^{-1}, which was greater than the combined production of the entire infaunal mollusc, and polychaete populations from the same area, which was 20.65 g dry wt m^{-2}.

A simple examination of the biomass of animals in an estuarine mudflat would tend to list the bivalve molluscs such as *Macoma* as the most important components of the ecosystem. When consideration is given to the productivity of the fauna the annelid worms, however, such as *Nereis, Ampharete*, and oligochaetes, with their high $P:B$ ratios (1.6–5.5) can often be seen to be more important.

4.2.3 Crustacea

The euryhaline amphipod *C. volutator* is an important component of many estuarine ecosystems. It lives within the upper 5 cm of the mud in a typical U-shaped burrow, emerging to collect fragments of detritus from the area around the burrow (Fig. 4.6).

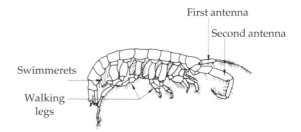

Figure 4.6 *C. volutator*. Adult female in side view. This animal lives in the surface layers on mudflats, where it inhabits U-shaped burrows, selectively feeding on detritus.

Table 4.4 Productivity (gC m^{-2} year^{-1}) of estuarine Crustacea

Species	Productivity (gC m^{-2} year^{-1})	Locality
C. volutator	47.8	Thames estuary/ Benfleet creek
C. volutator	9	Swedish west coast estuaries
Pontoporeia affinis	3.1	Baltic sea @ 46 m

Source: From various sources.

Corophium volutator is a selective deposit-feeder, which can discriminate between different kinds of sand or mud. Its field distribution is the result of the choice of suitable sediments coupled with the appropriate salinity regime (Table 4.4). It has been found that bacteria are more important than diatoms in the diet of *Corophium*, but *Corophium* can only utilize the bacteria that are adsorbed to clay and silt particles of 4–63 μm diameters. *Corophium* is unable to filter-feed on bacteria from the water, unless the appropriate size silt particles are also present. *Corophium* often coexists in sediments with the snail *H. ulvae*, but competition is minimized by the selection of particles for feeding of different sizes, with *Hydrobia* taking the larger particles, and *Corophium* the smaller. Seafoam, which may be stranded on intertidal areas at high tide has been shown to be a valuable food source for *Corophium*, when it occurs. The seafoam traps POM, such as algae, fungal spores, and plant fragments, all of which can be consumed by *Corophium*.

4.3 The surface dwellers—benthic suspension-feeders and algal grazers

4.3.1 Benthic suspension-feeders

The edible (or blue) mussel *M. edulis* develops large aggregated populations within estuaries (Fig. 4.7). The accumulation of dead and live mussels typically leads to a mussel bed community developing. In the Netherlands harvesting of mussels is a major industry. From a 140-km^2 Dutch estuary, for example, mussel harvesters collect an average annually of 19,225 tonnes of mussels.

For all suspension-feeders the highest levels of production are only maintained in the mussel or oyster beds, and average levels of production for the whole estuary are much lower. For example, the highest level of production of the *Mytilus* population of 268 g flesh dry wt m^{-2} year^{-1} in the Ythan is only maintained in the mussel beds of the estuary, which

occupy 18% of the intertidal area of the Ythan. The level of production of the *Mytilus* averaged over the whole estuary is 48.7 g flesh dry wt m^{-2} year^{-1}, which is more closely comparable to other estuarine fauna (Table 4.2).

Large populations of the American oyster, *Crassostrea*, are typical of the high salinity salt marsh tidal estuarine systems common to the southeastern United States, where they are economically important. In the North Inlet estuarine ecosystem near Georgetown, South Carolina, the oysters form extensive bars in places completely lining the intertidal creek banks. Oysters may have a valuable role to play in increasing the productivity of the area in which they live by filtering out food from the water column above them and depositing material as feces and pseudofeces, which is available to the benthic organisms. The Pacific oyster, *Crassostrea gigas*, can deposit material on the bottom, equal to that resulting from the combined activity of plankton and gravitational sedimentation in a 45-m column of water, and furthermore the biodeposited particles are

Figure 4.7 Mussel bed on the Ythan estuary, eastern Scotland. This suspension-feeding animal is often, as here, the main inhabitant of the lower reaches of the estuary, feeding mainly on material brought in from the sea on the incoming tide.

smaller and have a higher nutritive potential to browsing organisms. One square meter of oyster bed can filter 282,720 1 year^{-1}, and retain about 11% of the food material in the water, amounting to 2570 kcal m^{-2} year^{-1}. Much of the energy from the food filtered out from the water is utilized in metabolism and gamete production, and small amounts of energy contribute to shell growth, but the bulk of the energy ingested, amounting to 1545 kcal m^{-2} year^{-1} is rejected by the gills or passes through the gut undigested. This material is deposited on the bottom around the oysters where it can support a larger number of small deposit-feeders.

The edible cockle *Cerastoderma* (=*Cardium*) *edule* lives buried within the mud, with short incurrent and excurrent siphons protruding above the surface when they are covered by the tide. Its distribution in estuaries is generally patchy and production may vary annually dependent on the success of the spat-fall. The newly settled spat may achieve densities of 2400 m^{-2}, but high rates of mortality soon reduce that number.

Smaal and Prins (1993) have shown that suspension-feeders may generally be limited by the food supply, which is directly affected by the residence time of the water in the estuary. They found that the biomass of the suspension-feeders was highest in estuaries where the water was being exchanged rapidly, and hence the food supply was constantly being replenished. In general suspension-feeders are constrained by major hydrodynamic (=physical or mixing) constraints, and therefore can only occur patchily on a limited fraction of the total surface area of the estuary

4.3.2 Grazers

The winkle *Littorina littorea* is a grazer on algal films, utilizing its radula to remove this film from rocky outcrops or off the surface of mud and sand deposits. Within estuaries, *L. littorea* is very patchily distributed, and may often be found in salt marshes, eel-grass meadows and mussel beds.

In *Spartina* salt-marsh ecosystems, large quantities of plant fragments are potentially available to primary consumer animals. However, studies of the mud-snails in the salt marshes of Georgia have shown that littler of the *Spartina* detritus is utilized or available as an energy source for the snails; instead they rely on the benthic algae or the bacteria, which live on the plant fragments.

4.4 Meiofauna

The meiofauna are those animals, which are able to pass through a 0.5-mm (500 μm) sieve, but are retained by finer sieves (usually 62 μm). They are thus defined mainly by their body size, and include chiefly animals with small elongated bodies, which live interstitially in sand, or in the loose upper layers of mud. The most abundant of the meiofauna are the nematodes, and along with them may be found many groups of animals, especially harpacticoid copepods, and also Turbellaria, Gastrotricha, Tardigrada, Archiannelida, Coelenterata, and Annelida. Earlier studies of the meiofauna tended to consider their role in the energetics of marine and estuarine benthic ecosystems as being unimportant. Latterly there has been a realization that despite their small individual size and total biomass, the meiofauna can have a high productivity (high $P:B$ ratio) and may contribute substantially to the production of the estuarine benthos. The meiobenthic community has been found to be similar in many estuaries, being controlled more by predation and/or interference from macrofauna than by food supply.

A comprehensive survey of the meiofauna of the Forth estuary, Scotland, recorded a total of 172 different species, with a mean dry weight biomass for all the meiofauna of 1.1 g m^{-2}. Peak biomasses of about 4.0 g m^{-2} were recorded close to a sewage-works and an industrial effluent. Nematodes were found to be the numerically dominant taxon, supplemented by oligochaetes in the upper estuary, and copepods and polychaetes in the middle

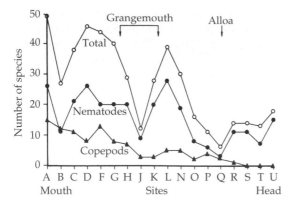

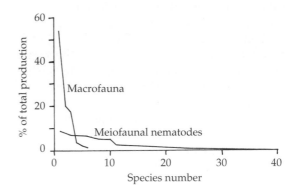

Figure 4.8 Meiofaunal species richness in duplicate 5.5 cm² cores taken from lower shore sites along the Forth estuary, Scotland. Note the direction plotted is the reverse of the data in Fig. 3.8. (From Moore 1987.)

Figure 4.9 Partitioning of annual production amongst macrofaunal species and meiofaunal nematode species from a mudflat in the Lynher estuary, England. (After Warwick 1981.)

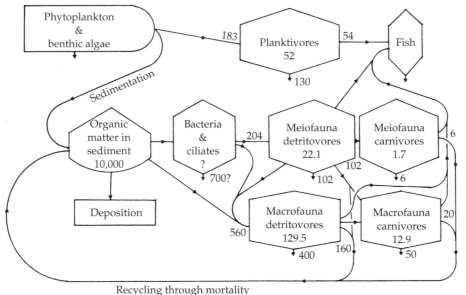

Figure 4.10 Energy flow model (Odum circuit language) of the benthic ecosystem of the brackish Asko/Landsort area of the Baltic Sea. Units are kilo joules per square meter per year. Figure to the left of the hexagons represent assimilation, figures within hexagons represent biomass, figures to the right of the hexagons represent production, and those below the hexagons represent respiration. (From Ankar and Elmgren 1976.)

and lower estuary. The total number of species was 50 per sample (2 × 5.5 cm²) at the mouth of the estuary, and decreased steadily within the estuary to below 10 in the upper estuary, increasing to over 15 at the freshwater head (Fig. 4.8). The dip in species number in the

Grangemouth area is apparently related to industrial discharges at that locality.

The most abundant representatives of the meiofauna in estuaries are the free-living nematode worms. Figure 4.9 shows the partitioning of annual production amongst

Table 4.5 Productivity (gC m^{-2} year^{-1}) of estuarine meiofauna

Group	Productivity (gC m^{-2} year^{-1})	Locality
Nematodes	6.6	Lynher, United Kingdom
All meiofauna	2.75	Asko Landsort, Baltic Sea
All meiofauna	20.39	Lynher, United Kingdom

Source: From various sources.

macrofauna, and the nematode component of the meiofauna at one estuarine site. These results suggest that for 40 species of meiofauna to coexist in 1 ml of sediment, the sedimentary environment must be highly structured on a microscopic scale, and that the organization of meiofaunal communities must be extremely complex. Estimates have been made of the contribution of the meiofauna to productivity of brackish ecosystems for sediment bottoms in the Askö-Landsort area of the Baltic (Fig. 4.10). The mean meiofaunal biomass here was 1.16 g dry wt m^{-2} with the largest contribution (0.4 g) from the ostracods, while the nematodes contributed only 24% of the biomass. Other groups represented were Turbellaria, Kinorhyncha, Harpacticoida, and Halacaridae. Most (91%) of the meiofauna was classified as detritivorous. The productivity of this and other meiofaunal communities is shown in Table 4.5.

Bacteria, microfauna, meiofauna, temporary meiofauna, and small macrofaunal elements in the benthos of estuaries and coastal areas can be regarded as one complex, characterized by the small size of the individuals, a high turnover rate ($P : B$ ratio), relatively short life spans, and a complicated trophic structure. This grouping has been referred to by Kuipers *et al.* (1981) as the "small food web"; they have shown that in the Wadden Sea these elements may consume 70–80% of all organic material available. The carbon flow through the intertidal of the Wadden Sea is summarized in Fig. 4.11. It can be seen that of the total carbon input of 350 gC m^{-2} year^{-1} (250 gC as detritus, 100 gC

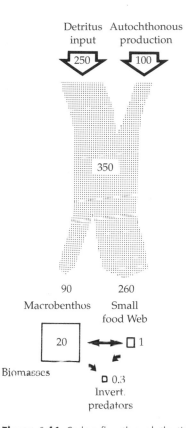

Figure 4.11 Carbon flow through the tidal flat ecosystem of the western Wadden Sea, The Netherlands. Main arrow shows the flow of detritus input and autochthonous production toward macrobenthos and the "small food web." Units as gram carbon per square meter per year. Boxes indicate biomasses of benthic faunal components, in units of gram carbon per square meter.

as primary production), only 90 gC m^{-2} year^{-1} is consumed by the macrobenthos, and 260 gC m^{-2} year^{-1} is consumed by the small food web. The mean biomasses are: macrobenthos (primary consumers) 20 gC m^{-2} macrobenthic predators (secondary consumers) 0.3 gC m^{-2}; and the small food web 1.0 gC m^{-2}. It may thus be seen that the macrobenthos and the small food web may compete for food, and this "biotic" factor may be the most critical limiting factor for the production of the macrobenthos, rather than any physical factor. The small food web does offer a substantial source of small food items for juvenile stages of secondary consumers such as shrimps, crabs, and fishes. This may serve to explain why so many carnivores from the open sea use shallow estuarine areas as nursery areas for their youngest stages, as discussed in the next chapter.

4.5 Zooplankton

The zooplankton is commonly defined as the animals that float in the water column, and have only limited swimming capacity. Few estimates of the biomass of the zooplankton in estuaries have been undertaken and all results are many times smaller than the rich benthic populations that we have already discussed. Estuarine zooplankton can be sub-divided either by size into micro, meso, and macrozooplankton, or by the duration of the zooplanktonic stage. The permanent members of the zooplankton, such as the calanoid copepods *Acartia* or *Eurytemora*, are known as the holoplankton, and the temporary members, such as the larvae of benthic forms (e.g. oysters or barnacles) whose life in the zooplankton is a dispersal phase, are known as the meroplankton.

The microzooplankton is most typically protozoans, such as flagellates and ciliates, which graze on bacteria and phytoplankton. Abundances are in the order of up to 10^9 m^{-3} as shown in Table 4.6. The impact of protozoan herbivory on the size of the bacterial populations can be very important, as shown in Table 4.7. It can be seen that up to 100% day^{-1} of bacterial biomass and production and up to 60% of primary production by phytoplankton can be grazed by protozoans in estuaries.

The mesozooplankton is most typically the calanoid copepods, including *Eurytemora*, *Acartia*, and *Pseudodiaptomus*, with densities of up to 100,000 Ind. m^{-3}, but usually less. The macrozooplankton includes the mysid shrimps (*Neomysis, Praunus*), also referred to as the hyperbenthos, as well as the large gelatinous zooplankton, such as comb jellies

Table 4.6 Estimates of abundance of microzooplankton (number m^{-3})—flagellates ciliates and bacteria in the pelagic zones of temperate estuaries

Location	Flagellates ($\times 10^9$)	Ciliates ($\times 10^6$)	Bacteria ($\times 10^{12}$)
Sapelo island, USA	1.3–31	0.08–0.09	—
Georgia, USA	2.3–4	16–29	2–17
Boston Harbor, USA	0.25	10–30	—
Hudson Estuary, USA	0.9–1.7	35	1.9–7.6
Clyde estuary, Scotland	0.5–50	0.001–50	7.5
Loch Striven, Scotland	0.5–50	0.001–10	2.9

Source: Heip *et al.* (1995) and references therein.

Table 4.7 Grazing impact of protistans on bacteria and algae in estuarine waters

Location	Protistans	% of bacterial production grazed daily	% of phytoplankton production grazed
Georgia, USA	Flagellates and Ciliates	50	—
Chesapeake Bay, USA	Flagellates	100–23	50–60
Essex, England	Community	100	—
Bothnian Sea, Sweden	Community	100	—
Hudson estuary, USA	Community	3–21	—
Boston Harbor, USA	Ciliates	15–90	—
Narragansett Bay, USA	Flagellates	100	16–26 (annual)

Sources: Heip *et al.* (1995) and references therein.

(Ctenophores, for example, *Beroe*) as well as true jellyfish (Scyphozoans, for example, *Aurelia*).

The mesozooplankton can be divided into three main groups. (a) Allochthonous fresh (e.g. *Eurytemora*) feeding on allochthonous material brought in to the estuary from the river; (b) Autochthonous (e.g. *Acartia*) feeding on phytoplankton produced within the estuary; and (c) Allochthonous marine (e.g. *Pseudocalanus*) feeding on allochthonous material brought into the estuary with the tides from the sea (Fig. 4.12). Estimates of estuarine copepod production are generally in the range of 8–30 gC m^{-2} year^{-1}, which is close to values reported from coastal marine areas (Table 4.8).

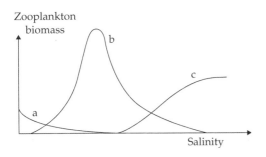

Figure 4.12 Idealized distribution of allochthonous and autochthonous zooplankton organisms along a salinity gradient. (a) Allochthonous fresh, (b) Autochthonous, and (c) Allochthonous marine. (After Heip *et al.* 1995.)

Table 4.8 Yearly production of mesozooplankton in estuarine waters compared to coastal marine waters (gC m^{-2} year^{-1})

Habitat	Location	Production (gC m^{-2} year^{-1})	$P : B$ (year^{-1})
Estuary	Patuxent River, USA	9	71
	Narragansett Bay, USA	18–23	140–150
	Westerschelde, Netherlands	11	45
Coastal marine	Northumberland, England	9–29	—
	Scotia shelf, Canada	8	9
	Southern North Sea, Dutch coast	12–23	

Source: Heip *et al.* (1995) and references therein.

High population densities of endemic estuarine zooplankton are often observed when primary production is small, showing that mesozooplankton are at least partly supported by detritus as food. Because of their ability to graze on non-algal diets the endemic mesozooplankton of estuaries may be able to maintain relatively high stocks all year round. Through selective feeding the mesozooplankton can greatly influence the pelagic system of estuaries, thus in Chesapeake Bay, it has been observed that on average less than 50% of the phytoplankton was consumed by the mesozooplankton, but on one occasion the entire primary production was grazed.

Two features potentially limit the zooplankton of estuaries. First by turbidity, which can limit phytoplankton production and thus limit the food available for the zooplankton, and second (and often more important) by currents which, particularly in small estuaries or those dominated by high river flow, can carry the members of the zooplankton out to sea. The main solution to these problems is for the zooplankton to stay near to the bottom and utilize as far as possible the inflowing marine currents, but these are tidally intermittent, and the best a species may achieve is to be carried to and fro about one location. Several species have been shown to undergo selective vertical migration upward and downward during different phases of the tide, by means of which they can maintain their location within the estuary.

The catches of zooplankton in an estuary can be variable. The most important factor affecting variability is the tide, and therefore to get accurate estimates of the zooplankton it is necessary to take samples over a 24-h period. Taking this into account, Lee and McAlice (1979) estimated that there were 23,882 copepods m^{-3} in the Damariscotta river estuary, Maine, with 8571 *Acartia tonsa* m^{-3}, 7360 *Eurytemora herdmanni* m^{-3} and 3753 *Acartia clausi* m^{-3}. The mean biomass of these populations of zooplankton was 0.086 g dry wt m^{-3}, which with a $P : B$ of 5 would only create an annual production of 0.43 g dry wt m^{-3}. Production of zooplankton has been measured in Chesapeake Bay, and the neighboring Patuxent river estuary, and suggests that zooplankton production is of the order of 5–10 gC m^{-2} year^{-1}, with the highest values in the upper estuarine reaches.

A detailed study of the zooplankton of the Forth estuary identified a total of 135 taxa, with 52 holoplanktonic taxa and 83 meroplanktonic taxa. The commonest holoplanktonic species were the calanoid copepods *Acartia longiremis*, *Acartia bifilosa inermis*, and *Centropages hamatus*, followed by the cyclopod copepod *Oithona similis* and the chaetognath *Sagitta elegans*. Larvae of *Littorina*, *Carcinas maenas*, and many polychaetes dominated the meroplankton. Within the estuary a clear sequence could be seen with the calanoid *Eurytemora affinis* totally dominating the upper reaches, a complex of *Acartia* species in the middle and lower reaches, with *Centropages*, *Oithona*, and *Pseudocalanus minutus* at the mouth, and penetrating further into the estuary in summer.

The greatest constraint on a pelagic population in an estuary is the lack of stability, resulting from tidal mixing and high river flows. Intuitively, a true planktonic fauna would seem unlikely to develop in the upper estuary where this instability is greatest. However, the copepod *E. affinis*, along with the hyperbenthos, such as the mysid *Neomysis integer*, thrives in such situations. Both these species may be regarded as epibenthic species, they both exhibit a marked shoaling behavior and they can apparently concentrate their distributions very close inshore at slack water. These behavioral features help them retain their position in the uppermost parts of the estuary, so that they can take advantage of a unique supply of food available there. This food supply is available at, and downstream of the freshwater–seawater interphase (FSI). The processes involved are summarized in Fig. 4.13. As river phytoplankton and other material enter the head of the estuary they undergo plasmolysis, or rupture, due to osmotic stress, which liberates both soluble and particulate organic material. These organic materials are

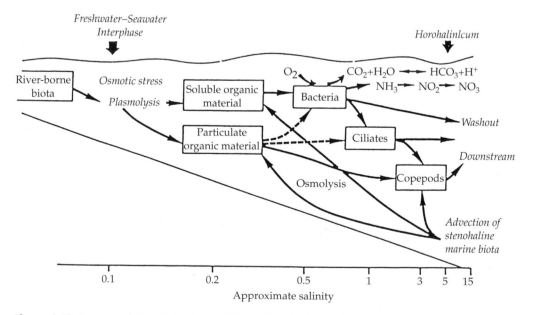

Figure 4.13 Summary of biological and related chemical mechanisms in low salinity regions of estuaries. From P. Burkhill. (pers. comm.)

utilized by bacteria. The bacteria are consumed by ciliate protozoa, which constitute the microzooplankton and are mainly non-tintinnid ciliates. Eurytemora then feeds upon the ciliates. Additional organic matter is carried into this region by inflowing bottom marine currents, and this material experiences osmolysis in the low salinity waters of the FSI. The abundant growth of bacteria in this zone, utilizing the organic material that is liberated here, will consume a considerable amount of oxygen, and may cause an "oxygen sag" in these waters. The concentration of zooplankton, albeit of only one species, may thus often be greater in these uppermost reaches of an estuary, than in any other part of the estuary.

Within the Patuxent estuary, Maryland, large quantities of detritus enter the water of the estuary from the adjacent salt marshes. This occurs particularly in the months of January and February, when ice scour carries the plant debris into the estuary. Within the estuary the mysid shrimp *Neomysis americana*, and also the planktonic copepod *E. affinis* use the detritus, presumably in a similar manner to that described above for the FSI. The input of the detritus in the late winter has been found to generate a substantial production of *Eurytemora* of 1.3–8.65 g dry wt m^{-2} year^{-1}. Fish larvae, especially those of the striped bass (*Roccus saxatilis*), consume the *Eurytemora* in turn. This simple food chain of detritus, to *Neomysis* and *Eurytemora*, to striped bass larvae is a rapid and efficient process, and it has proved to be possible to correlate the landings of striped bass with the severity of the winters, with cold winters causing extensive ice scour and export of detritus, leading to a large population of *Eurytemora* and an increased population of striped bass.

Oyster larvae are released into the zooplankton where they selectively swim and thus contribute actively to their retention within the estuary. This swimming behavior is connected to the increases in salinity that accompany a flood (incoming) tide, and thus helps maintain the oysters within the estuary

until they find a suitable settlement site. The role of estuaries in the transport of crab larvae has been investigated in the York river estuary, USA, where it was found that the larvae of crabs, which are restricted to estuaries (e.g. *Rhithropanopeus harrisii*) were adapted to be confined to the bottom waters of the estuary so that any transport that occurred would tend to be upstream and thus confine the animals to the estuary. Other more migratory crabs (e.g. *Callinectes sapidus*) produced larvae with no such adaptations so that the larvae tended to be carried out of the estuary into the sea, and later on the animals underwent a return immigration as juveniles or adults.

The estuarine hyperbenthos are a group of animals that generally live close to the bottom and perform vertical migrations into the water column. The commonest examples of estuarine hyperbenthos are the mysid shrimps, *Neomysis* and *Praunus* (Fig. 4.14). The mysids are considerably larger that the copepods and in some cases may be the main predators upon the mesozooplankton. Comparative studies of European estuaries have shown how similar is the pelagic community in all estuaries examined, and that the hyperbenthos (mysids etc.) in particular are an important link everywhere, feeding upon the smaller zooplankton and providing essential feeding for fish stocks.

Figure 4.14 A typical Mysid crustacean (*Opossum shrimp*). These animals can be some of the largest animals within the zooplankton of estuaries, being up to 3 cm long.

Other large members of the zooplankton are the comb-jellies (Ctenophore) and the true jellyfish (Scyphozoan). During the summer these gelatinous zooplankton can reach large numbers through apparent population explosions. Relatively little is known about these component of the estuarine fauna, however, clogging by jellyfish can be one of the commonest reasons for the cooling systems of estuarine located power stations being shut down.

4.6 The primary consumer community

Although large numbers of studies have been undertaken of the biomass and productivity of individual species within estuarine ecosystems, fewer studies have been made of entire estuarine trophic levels or communities. Overall estimates are compiled by adding the biomass and annual production for the different communities or zones within the estuary. In Table 4.9, as an example of such compilations, the biomass and production of the Spurn Bight area of the Humber estuary is given.

In Table 4.10 the total biomass and production of several estuaries are given. The levels of production for estuaries tabulated in Table 4.10 are much higher than reported for marine areas, where $4.5 \mathrm{~g} \mathrm{~m}^{-2} \mathrm{year}^{-1}$ has been recorded for Loch Ewe, Scotland, and $1.7 \mathrm{~g} \mathrm{~m}^{-2}$ year^{-1} has been recorded for the North Sea benthos, off Northumberland. The highest values recorded from freshwater localities are the studies of the eutrophic Loch Leven, Scotland, where the bottom community produced $525 \mathrm{~J} \mathrm{~m}^{-2} \mathrm{year}^{-1}$ (equivalent to 26.3 g ash-free dry wt $\mathrm{m}^{-2} \mathrm{year}^{-1}$).

It is the factors that control the distribution of the individual species, which are responsible for the assemblages that we observe in estuaries, rather than some process of biological organization. In assessing the total productivity of an estuarine ecosystem we should thus remember that while it is useful to group animals together in order to evaluate their contribution to the next trophic level, explanations of the reasons for the distribution of a

Table 4.9 Estimated biomass and production of four communities on the Spurn Bight mudflats of the Humber estuary in tonnes as an example of the calculation of total community production. Figures are given only for the main species. Figures in parenthesis in kilogram per hact

	Biomass (t)	Production (annual) (t)
Spartina zone (91 ha)		
H. ulvae	14.1	14.1[a]
Hediste diversicolor	8.7	22.2
Edukemius benedii	1.3	
Total for Zone	24.4 (268 kg ha^{-1})	36.4 (400 kg ha^{-1})
Upper shore SILT (332 ha)		
M. balthica	59.6	123.9
H. ulvae	13.1	13.1
H. diversicolor	12.7	42.9
Total for Zone	91.5 (276 kg ha^{-1})	182.3 (549 kg ha^{-1})
Midshore mud (1747 ha)		
M. balthica	321.0	229.0
C. edule	134.0	80.5
Nephtys spp.	64.5	147.0
H. diversicolor	25.3	83.4
H. ulvae	19.0	19.0
Retusa obtusa	5.1	24.0
Total for Zone	571.3 (327 kg ha^{-1})	582.9 (334 kg ha^{-1})
Downshore sand (786 ha)		
M. balthica	32.0	8.3
Nephtys spp.	9.3	21.3
C. edule	3.9	2.4
Total for Zone	46.9 (59.7 kg ha^{-1})	32.2 (41 kg ha^{-1})
Total for whole estuary (2956 ha)	734.1 (248 kg ha^{-1})	833.8 (282 kg ha^{-1})

[a] Productivity figures were not available for *Hydrobia*, this estimate assumes a $P:B$ ratio of l and is, therefore, likely to be a considerable underestimate. All these figures are liable to various degrees of under- or over-estimation and exclude *Mya* and mysids etc. In general, the upper shore sites are likely to be underestimated whereas the lower shore figures are more likely to be overestimates.

Sources: From Key (1983) in Jones (1988).

Table 4.10 Biomass and production values for so estuarine macrobenthic assemblages, all intertidal unless otherwise stated. Values as gram (ash-free dry weight) per square meter per year, gram carbon per square meter per year

Area	Biomass (g m^{-2})	Production (g m^{-2} year^{-1})	Biomass (gC m^{-2})	Production (gC m^{-2} year^{-1})
Long Island Sound, USA (subtidal)	54.6	21.4	27.3	10.7
Kiel Bight, Germany (shallow subtidal)	26.3	17.9	13.2	8.95
Baltic Sea, Sweden (shallow subtidal)			3.39	5.4
Lynher Estuary, England	13.0	13.3	5.43	5.46
Grevelingen estuary, Netherlands	20.8	50–57	10.5	25–29
Forth estuary, Scotland (upper estuary)			16.4	14.6
Forth estuary, Scotland (middle estuary)			5.1	6.7
Forth estuary, Scotland (lower estuary)			1.4	3.6
Forth estuary, Scotland (whole estuary)	10.5	12.9	5.25	6.45
Forth estuary, Scotland (subtidal)			0.7	0.6
Somme estuary, France			5.35	11.1
Berg River estuary, South Africa			9.5	44
Swartkops estuary, South Africa			34.7	38.9
Humber estuary, England	24.8	28.2		
Upper Waitema Harbour, New Zealand	17.9	27.3		

Soruces: Wilson (2002), Heip *et al.* (1995), and McLusky (1989).

particular species should generally be sought at the species level.

Studies of the benthic communities in several large European estuaries have shown a clear relationship to suspended matter concentrations and presumably silt deposition. In sediments covered by turbid waters communities dominated by deposit-feeders were found in the intertidal areas. In subtidal channels, tidal currents and the instability of the sediment became the limiting factors, leading to very poor communities (in terms of biomass,

abundance and diversity) in all estuaries examined. More "marine" macrobenthic communities are found in the seaward part of all estuaries, often dominated by filer feeders.

It has been demonstrated in this chapter that estuaries can sustain high levels of production, especially among the benthic primary consumers. These high levels of benthic production are responsible for maintaining abundant secondary consumers but at this stage we are led to ask how is it that estuaries can sustain high levels of benthic production. In Chapter 2, on primary producers, we stated that two distinct types of estuaries can be recognized. In the case of American estuaries, especially those studied in Louisiana, Florida, and Georgia, which have small tidal amplitudes, large quantities of detritus are derived from salt marshes, mangroves, or eel-grass beds within the estuary, which, supplemented by the primary production of phytobenthos and phytoplankton, produce large quantities of organic carbon, which are available to the primary consumers in the estuary. Currents from the estuary to the adjacent sea export any excess organic carbon. In the case of many north European estuaries, especially those studied in The Netherlands or the British Isles, which have large tidal amplitudes, much of the detritus that supports the benthos is derived by importing currents from the adjacent sea or river. Seventy-five percent of the particulate organic material in the Wadden Sea, for example, is derived from the North Sea, and only 25% is produced within the estuary by the phytoplankton or phytobenthos. The most distinctive difference is that in many American estuaries studied the marsh grass *Spartina alterniflora* occupies about 2/3 of the intertidal zone. Whereas in typical European estuaries salt-marsh plants are limited to a region above the height of high water of neap tides and 90–95% of the estuarine intertidal areas are bare mud and sandflats. Thus in the case of *Spartina*-dominated estuaries the surface-floating plant fragments will be carried seaward, whereas in mudflat-dominated estuaries the denser detrital fragments will be kept in suspension and carried into the estuary by the bottom inflowing saline currents.

Thus it can be recognized that the high level of primary production within estuaries is due to a supply of dissolved nutrients from various sources, especially by release from the sediments and by currents carrying nutrients from the sea and to a lesser extent rivers. The primary production of phytoplankton and phytobenthos is greatly supplemented by the addition of POM either from salt marshes or sea grasses, or from adjacent seas. Due to the

shallow nature of estuaries these sources of food become rapidly available to the benthic primary consumers. Sinking of surface production, coupled with tidal currents, may transport phytoplankton from the surface to the filter feeders within a matter of hours, and only a little longer time is required before the organic material sinks finally to the bottom where it becomes available to microbial decomposition and the detritus-feeding benthos.

From these considerations of the productivity of the primary consumer populations within estuaries it becomes clear that the high productivity is due to the large quantities of food material, which are potentially available. Abiotic factors, such as currents, tidal exposure, sediments, or salinity regimes may limit the distribution of individual species, and biotic factors, especially competition between the macrofauna and the small food web may limit the production of any one group of organisms. We have also seen in assessing the value of the primary consumers as food for the next trophic level, the secondary consumers or predators, that biomass is often not a good guide to the importance of a species. Many slow-growing bivalve molluscs have large biomasses, but even when the heavy shell is discarded, the remaining flesh may have a low productivity as expressed by a low $P : B$ ratio (or turnover rate). Other organisms such as annelid worms or amphipod crustaceans are faster-growing and shorter-lived and may often achieve production for the next trophic level that is over five times their biomass. The smallest organisms of all, within the small food web may have the least biomass, but the highest production. The importance of all the primary consumers as prey organisms will be discussed in the next chapter.

4.7 Role of primary consumers in the estuarine ecosystem

Initially many studies of estuaries concentrated on single species, measuring their consumption, growth, and the production of energy. In time such studies have been combined in order to build up food webs or trophic diagrams, and finally Baird and Ulanowicz (1993) show us how far the study of trophic structure, cycling, and the ecosystem properties of estuaries can develop. For the small Ythan estuary, Scotland they found that the main flows of energy were through the sediment pools and that all living components in the estuary were involved in the cycling process. For the much larger Ems estuary a pelagic (or water column) cycle structure was separate from the benthic cycle, with the benthic suspension-feeders as a link between the two cycles. This difference was due to an apparent lack of resuspension of sediment particulate organic carbon in the Ems. Baird and Ulanowicz (1993) also showed that the functioning of a "pristine" estuary was very similar to more anthropogenically enriched estuaries. These studies do show the crucial role of the benthos in estuaries as the principal energy link between the primary food supply and the consumers such as birds and fish. They also emphasize the need to analyze the role of benthos suspension-feeders separately from the deposit-feeders and the benthic carnivores.

Modeling of the Westerschelde estuary shows that 80% of the carbon that is respired is imported into the system (Soetaert and Herman 1995a), mostly from upstream (inland), but also from the sea, including the importation and subsequent death of marine zooplankton. Locally produced carbon is relatively unimportant. Due to the high heterotrophic activity (net heterotrophic means that respiration is greater than autotrophic production; i.e. a food web based on detritus), nearly all imported and locally produced carbon is lost in the estuary itself, and the estuary is an insignificant source of organic material to the adjacent sea. The estuary thus acts as a trap for reactive organic matter, both from the land, from the sea or produced within the estuary. Internal cycling, mainly in the water column, results in the removal of most of the carbon. All the imported, relatively reactive organic carbon is incorporated into the estuarine food

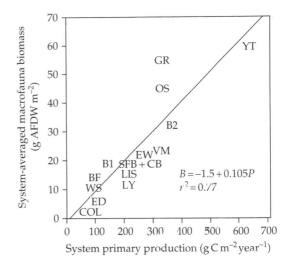

Figure 4.15 Relationship between system-averaged macrobenthic biomass and primary productivity of shallow well-mixed systems. (After Herman *et al.* 1999.)

web, and most of the organic matter that enters the estuary is used there. Only the more refractory (=difficult to break down) part escapes to the sea.

In a review of the ecology of estuarine macrobenthos, Herman *et al.* (1999) conclude that the coupling between benthic and pelagic systems is intense. Average benthic biomass is limited by primary productivity of the system. In Fig. 4.15 data on the relationship between benthic biomass and system productivity is assembled, where it can be seen that for these shallow well-mixed estuarine systems there is a clear and highly significant relationship between primary productivity and macrofaunal biomass. The relationship suggests that between 5% and 25% of the annual primary productivity is consumed by macrobenthos.

On an estuary-wide basis suspension-feeders can be the dominant component (with respect to biomass) of estuarine benthic assemblages, with values of 41–86% of biomass being due to suspension-feeders in several less-turbid estuaries. In more turbid estuaries, however, the deposit-feeders dominate the biomass. Suspension-feeders typically occur in much higher local biomass (e.g. *Mussels or Oysters* in "beds") compared to deposit-feeders, which are more evenly spread throughout the estuary at lower biomass levels. Thus at a large scale food may be the limiting factor for benthic biomass. Depending on the characteristics of the system, grazing by benthic suspension-feeders may be the most important factor determining dynamics of the entire estuarine system. The general conclusions are that a substantial fraction of the carbon flow in estuarine systems passes through the macrobenthic populations, and that therefore macrobenthic populations at an estuarine system level may be limited by food fluxes to the sediments.

The distribution of suspension feeders is critically controlled by the availability of food, and the seston within the body water as well as water currents. Grazing by benthic suspension-feeders has been shown effectively to control phytoplankton in estuaries such as the Oosterschelde in the Netherlands, and it has been shown that benthic grazing can limit phytoplankton bloom development in shallow water situations where vigorous vertical mixing is occurring and where grazing is sufficiently intense. But is has also been shown that as soon as even a mild form of permanent vertical stratification affects the system, then benthic grazing is no longer a major factor.

The distribution of deposit-feeders is controlled in large part by the nature of sediment, the flux of carbon into the sediment, and the availability of particulate organic carbon in the sediment, all of which are determined by complex processes, including biotic and abiotic factors. Hydrodynamic processes, such as the currents discussed in the first chapter, largely determine sediment texture or particle size. The organic content of the sediment depends both on deposition from the water column as well as the reworking of the sediment by its faunal inhabitants.

Deposit-feeders transport particles and fluid during feeding, burrowing, tube construction, and irrigation activity. By enhancing transport of particulate organic carbon to deeper sediment layers, these organisms stimulate anaerobic degradation processes and so affect the form and rate at which metabolites are returned to the water column. This bioturbation causes and upward movement of reduced components such as sulfide, as a consequence of which sedimentary oxygen uptake can increase. The release of nutrients such as nitrogen can be affected, and the reduction in the release of nitrogen due to benthic denitrification can directly influence the availability of this nutrient for phytoplankton production. The animals of the primary consumer community thus both influence the productivity of the primary producers, as well as providing food for the secondary consumers.

CHAPTER 5

The secondary consumers: carnivores

5.1 Introduction

The secondary consumers of estuaries are many and varied, the most conspicuous are the large numbers of birds, especially waders, gulls and wildfowl, which are attracted to estuaries as feeding areas. The birds mostly feed on the rich intertidal populations of annelids, crustaceans, and molluscs that are exposed by the tide. As the tide rises, many find the richest feeding at the water's edge. At high tide they may roost among salt marshes or else find other food in nearby fields if it is available. When the tide covers the mudflats large populations of fish, such as the flounder (*Platichthys flesus*) move onto the intertidal areas to avail themselves of the food supplies there. Along with the fish come the invertebrate predators, such as the shore crab (*Carcinus maenas*) or shrimps (*Crangon crangon*). Living buried within the mud may

be other carnivorous invertebrates, such as the cat-worm (*Nephtys hombergi*). All of these are attracted to estuaries by the large and productive populations of the primary consumers, which are dependent on plant and detritus production, which as we have seen are maintained by the ability of estuaries to trap nutrients and food particles. These rich food sources also make the estuaries less dependent on the seasonal fluctuations that characterize so many other temperate ecosystems. It is the productivity of the primary consumer that supports the variety of secondary consumers, many of which are temporary inhabitants of the estuary attracted by particular prey species at particular times of the year. A single prey species may be fed upon at different stages in its life by many different predators (Fig. 5.1).

In this chapter, we shall examine the impact of the various secondary consumers on the populations of primary consumers, and also

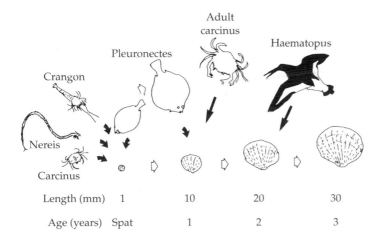

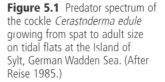

Figure 5.1 Predator spectrum of the cockle *Cerastoderma edule* growing from spat to adult size on tidal flats at the Island of Sylt, German Wadden Sea. (After Reise 1985.)

how the secondary consumers share or compete for the food supplies between themselves. Almost all the estuarine birds, fish, and crabs feed on the animals of the primary consumer trophic level, but it should be noted that a few birds by-pass this trophic level and instead feed directly on the primary producer plants. Most notable of these are the widgeon (*Anas penelope*) and Brent geese (*Branta bernicla*) feeding on eelgrasses and algae. Other birds such as the fish-eating birds (e.g. cormorant and merganser), or mammals such as seals, feed on the members of the secondary consumer trophic level, and should properly be called the tertiary consumers.

Migrations between habitats represent a fundamental aspect of the ecology of populations and individuals. The complex movements between breeding, wintering, or feeding areas during various life-history stages are often critical to the survival of species. Estuaries are viewed as critical nursery environments supporting a rich variety of seasonal visitors. The best-known examples are probably waterfowl, but estuaries are used in a similar way by fish and shrimps for feeding and wintering.

5.2 Fish

Fish populations in estuaries can be abundant with a wide diversity of species. Much of the abundance is seasonal as marine fish move into the estuary to breed and having used the estuary as a nursery the young fish grow and move out to sea again (Fig. 5.2). Other fish such as salmon (*Salmo salar*) and eels (*Anguilla anguilla*) use the estuary as a migratory route to get from rivers to the sea and vice versa, rarely feeding in the estuary. Only a few species live in the estuary throughout the year. Notable amongst these residents is the killifish (*Fundulus heteroclitus*) in American *Spartina*-dominated estuaries and various species of flatfish such as flounders (*P. flesus, Pseudopleuronectes americanus*) in mudflat-dominated estuaries. The flounders' food consumption may often rival or exceed that of the birds.

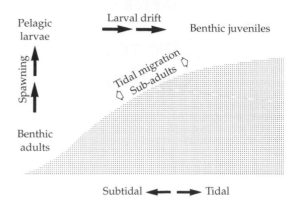

Figure 5.2 Common life cycle in demersal fish and mobile crustaceans living in estuarine and coastal regions. The pelagic larvae develop offshore, drifts inshore, and there become benthic juveniles. Later stages migrate tidally. The estuarine intertidal areas function as nursery grounds. (After Reise 1985.)

The fish fauna of estuaries may be classified into various categories (Box 5.1). The marine components are generally the dominant contributors to the diversity of estuarine species, here fish, and true estuarine, and other components contribute fewer species.

Estuaries perform a crucial role in the population dynamics of many fish. They provide a migratory route for anadromous and catadromous species and an environment for truly estuarine residents. Freshwater fish colonize the head of estuaries while typical marine species use the mouth. Many marine fish enter and remain in estuaries for a brief period of time, often in large numbers and particularly during the early part of their life cycle. A number of mechanisms have been suggested to explain this feature, including the avoidance or attraction to abiotic factors or gradients, the reduction intraspecific competition with adults of the same species, the reduction in predation levels and the increase in food availability.

Young fish utilize estuaries and near-shore marine areas as nursery areas in order to benefit from the availability of food and perhaps also to gain protection from predators. On the large estuary of the Patuxent in

Box 5.1 Ecological guilds of estuarine fishes (from Elliott and Dewailly 1995)			
Ecological guild	**Abbreviation**	**Definition**	**Example**
Diadromous species	CA	Catadromous or anadromous migrant species, which use the estuary to pass between salt and fresh waters for feeding and spawning.	Eel, Salmon, Trout, Sea lamprey
Freshwater species	FW	Species that occasionally enter brackish waters from freshwaters but have no apparent estuarine requirements.	Stickleback, Perch, Bream
Estuarine residents	ER	Truly estuarine resident species, which spend their entire lives in estuaries.	Flounder, Pogge, Eelpout
Marine adventitious species	MA	Visitors, who appear irregularly in the estuary, but have no apparent estuarine requirements.	Many marine fishes
Marine juveniles migrants	MJ	Species which use the estuary primarily as a nursery ground, usually spawning and spending much of their adult life at sea, but often returning seasonally to the estuary.	Herring, Cod, Haddock, Whiting, Plaice
Marine seasonal migrants	MS	Species that have regular seasonal visits to the estuary, usually as adults.	Sprat, Grey Mullet, Grey Gurnard

Maryland, for example, a clear seasonal pattern of abundance of fish has been noted with collections of fish more abundant and diverse in warmer water (spring and summer) and on the shores of the estuary. Collections in colder water (autumn and winter) are less abundant and diverse and mainly in the main channel of the estuary. These differences reflect the large number of young fish that use the shore areas as nursery feeding areas in spring and summer.

Estimates of the fish population in estuaries are more difficult to make than, for example, estimates of estuarine bird populations. Some idea of the diversity and abundance of estuarine fish populations can be gained from an examination of the large numbers of fish that become trapped on the cooling-water intake screens at power stations. At the Fawley power station in Southampton Water, England, for example, fish catches of up to 60,000 per week occur. Elsewhere in Britain a massive influx of sprats (*Sprattus sprattus*) blocking the cooling-water intake screens of a power station caused extensive damage. Collections at three power stations on the Scheldt estuary show both the numbers and variety of fish that enter estuaries (Fig. 5.3).

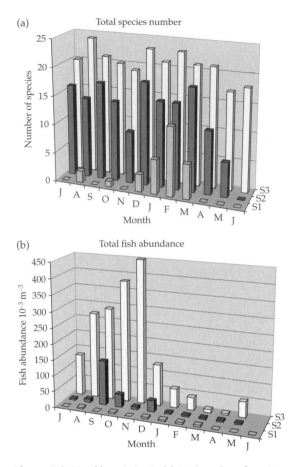

Figure 5.3 Monthly variation in (a) total number of species and (b) total fish abundance as recorded in monthly samples collected at the cooling water intakes of three power stations in the Scheldt estuary (Belgium/Netherlands). S1—salinity 0–1, S2—salinity 0.6–9.5, S3—salinity 3.4–13.4. (After Maes *et al.* 1998.)

The spawning areas of plaice (*Pleuronectes platessa*) in the southern part of the North Sea are 30–90 km off the Dutch coast. In the absence of adequate food supplies in the North Sea the very small 0-group plaice swim pelagically. When they reach the Wadden Sea the presence of adequate benthic food induces the young fish to settle on the bottom. The inward movement of young plaice into the Wadden Sea, to exploit the rich food available, is brought about by a selective use of the tidal currents. Within the Wadden Sea the young

plaice then consume as much as half of the invertebrate food available for all secondary consumers between March and July. It has been found that much of the food of the young plaice may be the siphons of the bivalve *Tellina tenuis*, and in any one year *Tellina* may have to regrow its siphons several times to cope with the fish predation. In the Netherlands, the tail tips of *Arenicola* form a substantial part of the fishes' diet, along with the siphons of *Macoma*.

Flatfish feeding in Dutch estuaries use at least two different feeding strategies. Plaice and flounder use tidal migration and feed only on the tidal flats during high tide. During low tide they wait along the edge of the tidal channel. Some individuals may swim 4 km or more per tide, but they are richly rewarded by the rich supply of prey items on the intertidal areas. Dab and sole, by contrast, are characterized by the absence of tidal migration, and remain continuously in subtidal areas feeding there. Table 5.1 compares the abundance of the major macrobenthic species at different levels of the Oosterschelde estuary intertidal areas, with the stomach contents of the plaice and flounder. In general the differences between the fish species, match the differences in the faunal distribution with plaice feeding on the low flats, and flounder on the high and middle flats.

Fifty-four species of fish have been recorded from the small Slocum river estuary, Massachusetts over a 2-year period, with five species dominant—the mummichog or killifish, *F. heteroclitus*, Atlantic silverside, *Menidia menidia*, four-spine stickleback, *Apeltes quadracus*, striped killifish, *Fundulus majalis* and sheepsheaf minnow, *Cyprinodor variegatus*. These five species dominated the estuary throughout the year, with the main summer additions being species, which used the estuary as a nursery ground, such as Atlantic menhaden, *Brevoortia tyrannus* and white perch, *Morone americana*. The main factor that affects the abundance and diversity of the fish populations is temperature, with maximum diversity of fish in the warmest month, July, and the least

Table 5.1 Occurrence of seven macrobenthic species at three tidal levels on the tidal flats of the Oosterschelde estuary, compared to the average number of prey specimens in the stomach and guts of plaice and flounder

	Density of prey species (mean m^{-2})			Average number of prey (mean per fish)	
	High flats	Middle flats	Low flats	Flounder	Plaice
Anaitides maculata	0.0	0.0	6.7	0.0	3.1
N. hombergii	0.0	11.1	4.4	0.1	0.8
Pygospio elegans	100.4	82.2	20.0	2.1	0.1
Arenicola marina—small	22.2	0.0	0.0	0.2	0.2
A. marina—large	26.7	6.8	55.6		
A. marina—tails	—	—	—	12.1	32.8
Lanice conchilega	0.0	0.0	4.4	0.1	0.8
H. ulvae	4575	7055	2.2	16.7	2.4
R. obtusa	8.9	0.0	2.2	0.0	3.4

Source: Wolff et al. (1981).

diversity in the coldest month, February. It was also found that areas with the maximal salinity variation had the most diverse fish faunas.

The killifish or mummichog, *F. heteroclitus* is one of the most abundant resident fishes within the tidal creeks of *Spartina*-dominated salt marshes where they feed principally on benthic invertebrates. Studies in the salt marshes of Delaware and Massachusetts have revealed annual productivity values of 10–16 g dry wt m^{-2} year^{-1}. The productivity values from Delaware were calculated on the basis of the populations of *Fundulus* within a 36-m long portion of a 3-km long creek. Within the 36-m portion the mean biomass was 3259–7298 g-wet wt, and the population numbered 378–814 individual fish. The Massachusetts study was based on the results derived from eight tidal creeks, in which minnow traps were placed, and the fish caught were marked, released, and then a proportion caught again. Such estimates of production suffer from possible errors in the estimation of the size of mobile populations, nevertheless, the consistency between the two studies does point to the high productivity of these fish populations. In non-estuarine studies of fish populations, productivity values of 0.5–4 g dry wt m^{-2} year^{-1} are typical so it can be seen that these studies of *Fundulus* are between 4 and 20 times higher. The killifish in Massachusetts have a 3-year life span, during which time they are subject to intense predation by birds and other larger fish. In contrast to many other estuarine fish species that are seasonally migratory, the *Fundulus* remain for their whole life confined to a single marsh system, apparently relying on the tides to bring their rich food supply to them. However, their sedentary behavior has permitted a close examination of their productivity to be made, in contrast to the many estuarine fish populations elsewhere about which little is often known.

On mudflat-dominated estuaries the most common fish are often the flounder (*P. flesus*) and the sand goby (*Gobius minutes*). On the Ythan estuary, Scotland, for example, the number of flounders varies from 0.05 m^{-2} in winter to 0.24 m^{-2} in July, representing biomasses of from 5 to 35 g wet wt m^{-2}. The total food consumption of flounders in the Ythan has been calculated as 58.1 kcal m^{-2} year^{-1}, which may be compared to the consumption of gobies at 8 kcal m^{-2} year^{-1}, and the combined consumption of birds (Oystercatcher, Dunlin,

Redshank, and Shelduck) in the same area at 23.9 kcal m^{-2} year^{-1}. Thus in this estuary at least the fish populations appear to consume three times the amount of food consumed by the birds.

The winter flounder (*P. americanus*) in Newfoundland feed mainly during the summer, consuming principally polychaetes, plant material, molluscs, capelin eggs, and fish remains. The winter flounder is the dominant fish in the polluted Mystic River estuary in Massachusetts, comprising 89% of the total fish biomass. The winter flounder goes through its entire cycle in the estuary feeding principally on the *Capitella* worms that are the dominant members of the benthic fauna. Only two other fish species spent the entire year in this estuary, the alewife (*Alosa pseudoharengus*) and smelt (*Osmerus mordax*). Another 18 species visited the estuary seasonally, mostly in summer. The biomass of fish for this estuary averaged 2 g wet wt m^{-2}, a low value that reflects the polluted conditions in this estuary.

A study of the fish assemblages in estuarine waters around Sweden has shown that the number of species is limited by the salinity, with fewer species in less saline waters, but the abundance of fish was determined by temperature, with more fish in warmer waters. The fish coexist with mobile invertebrates, such as crabs (*C. maenas*) and shrimps (*Crangon, Palaemon*), which feed on the same bottom fauna. In summertime the small fish have to compete with shrimps as well as other fish species for their food, and this may serve to limit the numbers of fish in an estuary. Apart from competing with the shrimps and other mobile fauna for benthic food, fish such as cod and flounder also eat these animals and in a study in one shallow bay it has been shown that the fish consumed about 7% of the production of the mobile epifauna.

The estuarine fish communities of the North Sea comprise several components each with differing requirements in the ecosystem. Some fish use the estuary during their whole life cycle, but most use it only during some seasons or for different stages of their life cycle. Many of these species make considerable feeding and spawning migrations, such as Herring, which drift from their spawning areas on the East of Scotland and England into the local estuaries, as well as across to the German Bight and Netherlands–Germany–Denmark coasts. For many, if not most, of the gadoids (Whiting, Saithe etc.) the older age groups occur in deeper waters, while the juveniles occur in coastal waters and estuaries. The estuaries of the North Sea are firmly recognized as important "nursery" areas for North Sea fish stocks. The process of selective tidal stream transport by larvae of the Flounder (*P. flesus*) encourages the retention of the larvae inside the estuary as a nursery ground. This is shown by the vertical movement of the larvae within the estuary deviating from the movement of suspended matter, so that the larvae could utilize tidal streams to remain within the estuary.

Plaice, Sole, and Dab are classic examples of species with nursery grounds. Their bottom-living 0-group stages occur exclusively in very shallow water (littoral or sub-littoral). These three species are therefore relatively scarce in the northern North Sea, where there is a dearth of suitable nursery grounds, while they are abundant round the perimeter of the southern North Sea, especially in the Wadden Sea and other North Sea estuaries. Dabs spend a comparatively short time in the nursery areas, but Plaice and Sole may spend 1 or 2 years. Cod too spend time inshore and in estuaries, with the German Bight being a major nursery. It is probable that a large proportion of the cod in the northern North Sea spend their first 1 or 2 years inshore along the UK coast, where they cannot be caught by conventional survey gear.

Reviews of the possible physiological reasons for estuarine usage by fish conclude that the most likely cause is the presence of high-density prey and hence favorable feeding areas. For the Forth they show that the estuary supports a population of about 0.5% of the

North Sea stocks of similar sized Plaice and Cod. The figure of 0.5% may appear small, but it must be remembered that the Forth estuary is much smaller than 0.5% of the area of the North Sea, so therefore it must be considered that its usage is much higher than expected on a simple proportional basis.

Three percent of North Sea Plaice utilize the outer Humber estuary as a nursery feeding ground. Studies into the year-class strength of Plaice in the North Sea have suggested that processes acting during their post-larval demersal stage in the nursery areas such as estuaries or the Wadden Sea may determine this. Year-class strength was in large part controlled by the influence of the predator, *C. crangon*, which feeds upon the Plaice when they are in these nurseries. For other fish, *Crangon* is essential as their principal dietary item, as shown by stomach analyses of juvenile Whiting, Cod, and Haddock found in estuaries.

The species richness of fish in tropical estuaries is greatest in the Indo-West Pacific, but larger systems usually have more species than smaller ones. Deep open water channels in the larger systems favor more of the larger species, particularly carangids and sharks, in addition to a higher number of occasional visitors. In addition, larger systems usually have a greater diversity of habitats. Hence comparisons of species richness between estuaries of the Indo-West Pacific and the East and West Tropical Atlantic should be viewed with caution. However, taking into account the large size of some Atlantic systems (e.g. Orinoco estuary, Terminós and Lagos Lagoons), it is apparent that even small estuaries in the Indo-West Pacific are usually more species rich (Table 5.2). Comparisons of overall fish biomasses and productivity (Tables 5.3 and 5.4) in different tropical estuaries are difficult because there are few published figures, and also because many of the results may not be comparable due to differences in methods. However, the scarce data suggest that fish biomass does not usually exceed 30 g m^{-2} and is most commonly between about 5 and 15 g m^{-2}. These values are an order of magnitude higher than those of tropical non-estuarine sea grass areas in shallow waters are (Blaber 1997).

The number of fish species in subtropical and tropical estuaries is much greater than in temperate regions. Where individual temperate estuaries may have about 20 species (Elliott and Hemingway 2002) and warm temperate ones about 50 (Lenanton and Hodgkin 1985), most medium to large subtropical and tropical estuarine areas have at least 100, with some reaching over 200 (Table 5.2). Additionally, the number of species in adjacent estuarine coastal areas often exceeds 300, but depends upon the presence of specific habitats such as sea grass beds. The composition of the fish faunas results from the interplay of a whole range of factors among which the most important are:

(a) Estuary size, depth, and physical regimes, particularly salinity and turbidity, as well as the types of habitats, particularly the composition and extent of mangroves.
(b) The nature and depth of adjacent marine waters and to a much lesser extent, freshwaters.
(c) The geographical location of the estuary, both in terms of latitude and in relation to marine features such as ocean currents, canyons, and reefs.

The classification of estuarine fishes has been almost as numerous as the classification of estuaries (Blaber 2002) with taxonomic, physiological, and ecological groupings based on attributes such as salinity tolerance, breeding, feeding, and migratory habits. Most research on estuarine fishes has been in temperate and warm temperate regions of Europe and North America where salinity has long been regarded as a key factor regulating the composition of estuarine fish faunas. The lower salinity of estuarine waters is considered to be the outstanding difference between estuarine waters and seawater. While this phenomenon has been amply demonstrated for warm temperate and temperate areas it is not at all clear in the

Table 5.2 Numbers of species recorded from different types of estuaries (O = open, B = blind, CL = coastal lake) in the four major zoogeographic regions

System	Country	Types and sizes	No. of species
Indo-West Pacific			
Alligator Creek	Australia	O small	150
Trinity	Australia	O medium	91
Embley	Australia	O large	197
Vellar Coleroon	India	O large	195
Chilka	India	CL large	152
Hooghly	India	O large	172
Chuwei	Taiwan	O small	30
Ponggol	Singapore	O medium	78
Matang	W. Malaysia	O medium	117
Kretam Kechil	E. Malaysia	O small	44
Purari	New Guinea	O large	140
Pagbilao	Philippines	O medium	128
Solomon Islands	Solomon Islands[a]	O small	136[a]
Morrumbene	Mocambique	O medium	113
Tudor Creek	Kenya	O small	83
Gazi Bay	Kenya	CL small	128
Kosi	South Africa	CL large	163
St Lucia	South Africa	CL large	110
Mhlanga	South Africa	B small	47
East Atlantic			
Lagos	Nigeria	CL large	79
Fatala	Guinea	O medium	102
Ebrié	Ivory Coast	CL large	153
Niger	Nigeria	O large	52
West Atlantic			
Guaratuba	Brazil	CL medium	61
Itamaraca	Brazil	O large	81
Orinoco	Venezuela	O large	87
Terminos	Mexico	CL large	122
Cienaga Grande	Colombia	CL large	114
Tortuguero	Costa Rica	O small	70
Sinnamary	French Guyana	O medium	83
East Pacific			
Rio Claro	Costa Rica	O small	22
Guerrero Lakes	Mexico[b]	CL small	105

[a] Incorporates 13 small estuaries—no one system more than 50 species.
[b] Sum of species for 10 small coastal lakes.
Source: Modified from Blaber (2000).

subtropics and tropics. Salinity is only one of an array of factors that determine which species are found in any one subtropical or tropical estuary. Salinity tolerance may play a role in the distribution of fishes in tropical estuaries, but as these systems often undergo very great fluctuations in salinity (often from almost 0 to 35), both diurnally and seasonally.

Table 5.3 Biomasses of fishes from various subtropical and tropical estuaries

Country	Faunal region	Estuary or habitat	Fish biomass (g m^{-2})
Malaysia	Indo-West Pacific	Matang	0.25–3.1
Australia		Alligator Creek	2.5–29.0
		Embley	5.0–16.0
		Albatross Bay (inshore)	5.0
		Albatross Bay (offshore)	12.0–30.0
Solomon Islands		Small estuaries	11.6
Florida, USA	Tropical West Atlantic	Mangrove estuaries	15.0
Mexico		Terminós Lagoon	0.4–3.4
Mexico	Tropical East Pacific	Mangrove-lined canal	7.9–12.5

Source: Modified from Blaber (2000).

Table 5.4 Production of fish in different types of tropical estuaries

Country	Estuary	Type	Tonnes (km^{-2})
India	Hooghly-Matlah	Open estuary system	11.4
	Vellar-Coloroon	Open estuary system	11.1
Malaysia	Larut-Matang	Open estuary system	38.64[a]
United States	Texas bays	Estuarine coastal waters	12.1
El Salvador	Jiquilisco	Estuarine coastal waters	1.7
Philippines	San Miguel Bay	Estuarine coastal waters	23.8[b]
South Africa	Kosi system	Coastal lakes and open estuary	1.0
Ivory Coast	Ébrié Lagoon	Coastal lake	16.0
India	Lake Chilka	Coastal lake	3.7
	Lake Pulicat	Coastal lake	2.6
	Mandapam Lagoon	Coastal lake	5.6
Ghana	Sakumo Lagoon	Coastal lake	15.0
Malagasy	Pangalanes Lagoon	Coastal lake	3.7
Colombia	Cienaga Grande	Coastal lake	12.0
Mexico	Caimanero Lagoon	Coastal lake	34.5
	Terminos lagoon	Coastal lake	20.0
	Tamiahua Lagoon	Coastal lake	4.7
Venezuela	Lake Maracaibo	Coastal lake	1.9
	Tacarigua Lagoon	Coastal lake	11.0

[a] Includes penaeid prawn catches and not coastal waters.
[b] Probably an overestimate as trawlable biomass only 2.13 tonnes km^{-2}.

Source: Modified from Blaber (2000).

A considerable degree of euryhalinity is a pre-condition and prerequisite for their inhabitants, as well as for many species in marine areas that suffer seasonal reductions or fluctuations in salinity.

For this reason, a grouping of tropical estuarine fishes based mainly on their salinity tolerances is not very useful or practical—especially as there has been little experimental work conducted on the salinity tolerances of most species. Based on their occurrence in low salinity waters almost all fit into a similar, very euryhaline category. This has led to a number of classification systems for subtropical and tropical estuarine fishes (Blaber 1997; Whitfield 1998), based mainly on how they utilize estuaries and where they spawn. The following classification scheme is similar, but gives increased importance to the "estuarine" category in the tropics due to recent advances in knowledge of spawning behavior (Blaber 2000).

1. Estuarine species that can complete their entire life cycle in the estuary. In the Subtropics they represent a small but significant part of the estuarine fish fauna. In the Tropics they represent a large part of the estuarine fish fauna, particularly in West Africa, South East Asia, and tropical South America.

2. Marine migrants from the sea—usually characterized as marine spawners. This is a large group in subtropical and tropical estuaries where they may occur both as adults and or only as adults. However, it is important to note that the division of this category from that of (1) above is often difficult in the tropics, because many of the species that fall into the marine migrant category in the subtropics, form part of the "estuarine" category (above) in the tropics.

3. Anadromous species that breed in freshwater and on their way to and from their spawning grounds spend time in estuaries. This group is important in temperate estuaries, but is a rare category in subtropical and tropical estuaries.

4. Freshwater migrants that move varying distances down estuaries, sometimes to spawn (e.g. eleotrids), but usually return to freshwater to breed. In the tropics this group is best represented in South and Central American estuaries where a number of pimelodid catfish, poeciliids, and characids penetrate estuarine waters. The freshwater cichlid *Oreochromis mossambicus* and the archer fishes (Toxotidae) are well known African and Southeast Asian examples.

The fish communities of subtropical and tropical estuaries of all four major tropical zoogeographic regions (Indo-West Pacific, East Atlantic, West Atlantic, and East Pacific) have many common characteristics. In almost all cases, fishes of marine origin dominate them with more than half the number of species as well as the number of individuals, being contributed by either fully estuarine species or marine migrants. Given the "estuarization" of many tropical coastal areas, this is not surprising and perhaps the term "marine" for many of these species is a misnomer, for many of them do not occur outside of estuarine coastal areas. The dominant taxa in each region are broadly similar, but there are some interesting contrasts. In all regions except the Indo-West Pacific, Sciaenidae are one of the dominant families. In the Indo-West Pacific sciaenids are important in the equatorial regions of South East Asia, but much less so elsewhere. This pattern may be connected with the amount of rainfall and the degree of estuarization of coastal waters. In East Africa, for example, there are few estuarine coastal waters and sciaenids, and although present they are not a dominant component of the fauna. Except for the sciaenids, the other dominant families are broadly comparable across all regions. It is noteworthy, however, that within the Indo-West Pacific, clupeids and engraulids are much more diverse and numerous in the estuarine coastal waters of equatorial South East Asia than in other areas.

5.3 Invertebrates

Invertebrate predators are principally crabs and shrimps, and carnivorous polychaetes, the former mobile with the tides, and the latter buried in the mud. Shore crabs (*C. maenas*) in the English estuaries show a seasonal migration. In the wintertime they are spread throughout the estuary, but during the summer the crabs move upstream and the lower half of the estuary becomes devoid of crabs. In November they again spread throughout the estuary. This upstream estuarine migration is apparently similar to the onshore summer migration that takes place in marine crabs. In the Isefjord, Denmark, a negative estuary, the shore crab (*C. maenas*) move inshore in May as the temperature rises above 9 °C, with the largest ones first, followed by the smaller individuals. They feed actively inshore during the summer and autumn and move offshore in October–November and hide more or less inactively in deeper water during the winter. Such migrations seem to be a clear adaptation to make use of the rich benthic production available during the summer months.

Caging experiments designed to exclude crabs and gobies, and monitor the effects of these species on benthic infauna, have shown that predation by the shore crab *C. maenas* can significantly reduce the abundance of small annelids on an estuarine mudflat, particularly the dominant polychaete *Manayunkia aestuarina*. Predation could be responsible for year-to-year variations in abundance of this species, whereas in this experiment, the fish had little direct effect on the abundance of prey.

In the Lynher estuary, Devon, the carnivorous benthic polychaete, *N. hombergi* occurs abundantly with a mean annual biomass of 3.947 g m^{-2} and an annual production of 7.335 g m^{-2} year^{-1}. Since populations of other macrofaunal invertebrates were inadequate to support this level of production, it is suggested that much of the food of *Nephtys* must be meiofaunal. The carnivorous polychaete, *Nereis virens* often forms substantial populations, feeding on smaller polychaetes. The ragworm *Nereis diversicolor* has been shown to be a predator on *Corophium volutator*, small *Macoma balthica* and chironomid larvae, as well as an omnivorous feeder on detritus.

Predation also occurs within the zooplankton, for example the planktonic sea gooseberry, the ctenophore, *Mnemiopsis leidyi*, which is a carnivore feeding on smaller members of the estuarine zooplankton, can reach large biomasses through its ability to rapidly increase

in population size as food becomes available. In Narragansett Bay, United States, a biomass peak of 60 g wet wt m^{-2} of *Mnemiopsis* has been reported.

The mobile shrimps and prawns (e.g. *C. crangon*) and the opossum shrimps (e.g. *Praunus flexuosus* and *Neomysis integer*) are often important predators in estuaries. The high densities of shrimps may often make them one of the main commercial catches in estuaries. Like crabs, shrimps tend to migrate further into estuaries and into shallower water in summertime and retreat to deeper water in wintertime. When feeding in estuaries they tend to be omnivorous, utilizing plant fragments as well as smaller planktonic animals and members of the benthos. In the Dutch Wadden Sea the juvenile *Crangon* feed on planktonic copepods, but as they grow the *Crangon* switch to feeding on benthic foraminifers and polychaetes. A study of the predators feeding on *Corophium* in a Swedish estuary, has shown that 98% of the annual production of *Corophium* is consumed by shrimps, crabs, and fish, with 57% being consumed by *Crangon*, 19% by *Carcinus*, and the balance (22%) divided between three fish species (Fig. 5.4). *Crangon* also feeds on *Nereis*

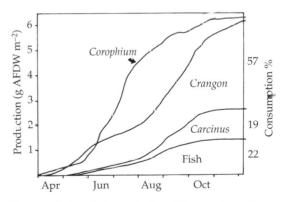

Figure 5.4 Cumulative production of *C. volutator*, and cumulative consumption of *Corophium* by the dominant epifaunal predators, in shallow areas of Gullmarsfjord on the Swedish West Coast. Percentage consumption indicated on right. Production given as g ash-free dry weight m^{-2}. (After Pihl 1985.)

and young bivalves, as well as juveniles of *Carcinus* and its own offspring. From several studies it has been concluded that *Crangon* is one of the major carnivores of shallow marine and estuarine areas, and exerts a major impact on the infaunal community.

Gizzard contents of the predatory snail *Retusa obtusa* show that the youngest and oldest feed on foraminifers, but the growing intermediate stages eat the snail *Hydrobia ulvae*. There is a close match of predator and prey size, so that young *Retusa* feed on small *Hydrobia*, and the sizes increase correspondingly throughout the summer of growth.

The total impact of invertebrate and fish predation in estuaries is large, especially as many of the predatory species move into the estuary in summertime when the maximal biomass of prey is available.

5.4 Birds

The numbers and variety of birds of estuaries vary throughout the year. In Europe, many of the most abundant species of Waders arrive in the autumn on migration from summer breeding habitats further north or east. Some of these stay in the estuaries of Britain and the Netherlands throughout the winter, while others move on south to overwinter in Africa.

Some of the ducks such as Shelduck or eider breed in the estuaries around the North Sea in summer, but they are supplemented in winter by others migrating from further north. Geese may be present in large numbers in estuaries as they pass through on migration while the Brent geese (*B. bernicla*) may be resident in the winter. Living throughout the year in the temperate estuaries of Europe are birds such as mallard, gulls, and cormorants (Fig. 5.5).

The bird predators of estuaries are highly mobile and typically exhibit clear tidal rhythms of activity associated with water movements and the activity of the prey in relation to the tides. The Redshank (*Tringa totanus*) and the Shelduck (*Tadorna tadorna*), which feed on *Macoma*, *Hydrobia* and other infaunal invertebrates feed mainly on the intertidal zone at low water, tending to be near the water's edge. Eider ducks (*Somateria m. mollissima*) feed in the shallow water at low tide, collecting in particular the mussel, *Mytilus edulis*. Others, such as the diving ducks, (goldeneye, *Bucephala clangula* or scaup, *Aythya marila*), along with cormorants (*Phalacrocorax carbo*) and mergansers (*Mergus serrator*) feed at high water by diving for their prey. In addition to the birds, which feed in an estuary others such as geese, various species of gulls, certain ducks, and swans may use an estuary primarily as a place

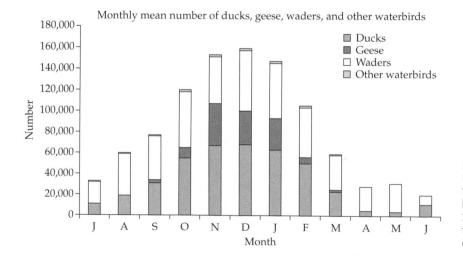

Figure 5.5 Monthly mean numbers of ducks, geese, waders, and other water birds counted in the Scheldt estuary during the period 1991–1997. (After Ysebaert *et al.* 2000.)

to roost and obtain their food either at sea or from land sites.

Many shore birds have adapted their food-searching behavior to suit the tidal and seasonal rhythms of their preferred prey. It has been generally shown that the bill-length of particular species of birds is suited to particular species of prey organism, but more subtle behavioral adaptations also occur. The bar-tailed godwit (*Limosa lapponica*) feeds principally on the lugworm *Arenicola marina*, and waits and watches for new *Arenicola* casts so that it can catch the lugworms as they defecate. The Grey plover (*Pluvialis squatarola*) searches by sight for its worm prey, relying on the movement of the prey (the ragworm, *N. diversicolor*) at the surface of the substratum. Knot (*Calidris canutus*) on the Tees estuary feed on *Hydrobia*, and like Shelduck on the Forth (see below), rely on the re-emergence of *Hydrobia* as the mudflats are reimmersed by the tide. The ragworm *N. diversicolor* retreats into its burrow as the tide ebbs off intertidal mudflats, but is nevertheless caught by curlew (*Numenius arquata*) relying on its sense of touch. Curlew also feed on shore crabs (*C. maenas*), but as these move offshore in winter, they are only important in the summer diet.

The invertebrate prey may adjust their behavior in response to the activity of the predators. *Macoma*, for example, in the Wash apparently remain close to the mud surface at low water whenever their predators are absent, but bury deeper into the substrate at low waters whenever predators are present. The position of burrowing animals within the substrate is most closely linked to the state of the tide. In experimental studies of Wadden Sea animals, it has been clearly shown that *H. ulvae* and *N. diversicolor* migrate from being buried in the mud to the surface in direct response to the inundations of the tide over the mudflats. There is no need to invoke any inherent rhythm to the animals' behavior, and thus the animals only come to the surface to feed just prior to inundation and when they are actually covered by the tide.

The waders are one of the most important groups of estuarine birds, especially in autumn and winter. They feed on a variety of prey organisms depending on what is available in their particular estuary. The principal prey of Waders in the Wash, eastern England is shown in Table 5.5, in which it can be seen how the prey resources have been divided between the species of Waders with only limited overlap between species.

Oystercatchers (*Haematopus ostralegus*) feed on mussel beds when they are completely exposed at low tide, by hammering on one of the bivalve shells until it splits open, or inserting their bill between the valves to cut the adductor muscle. Once the shell is open, the flesh is picked out and the shell discarded. In the Wash the Oystercatchers eat cockles and their daily consumption of cockles has been estimated as 181 cockles for each bird each day (=57.5 g total wet weight). Oystercatchers in North Wales feed on

Table 5.5 Principal prey species of the main wading birds in the Wash

Bird	Prey
Oystercatcher (*H. ostralegus*)	*C. edule* *M. edulis*
Knot (*C. canutus*)	*M. balthica* *C. edule*
Dunlin (*C. alpina*)	*H. ulvae* *N. diversicolor*
Redshank (*T. totanus*)	*C. maenas, Crangon* sp. *H. ulvae, Nereis* sp.
Bar-tailed godwit (*L. lapponica*)	*L. conchilega* *Nereis* sp., *M. balthica*
Turnstone (*Arenaria interpres*)	*C. edule* Among mussel beds
Grey plover (*P. squatarola*)	*L. conchilega* and various
Curlew (*Numenius arquata*)	*C. maenas,* *L. conchilega,* *A. marina*
Ringed plover (*Charadrius hiaticula*)	Not recorded
Sanderling (*Crocethia alba*)	Not recorded

Source: Goss-Custard *et al.* (1997).

the buried bivalve *Scrobicularia plana* by thrusting their bill deep into the mud, almost up to their eyes, then jerking out the bivalve. The extracted bivalve is then wedged apart by the bill, and the soft parts removed.

Within the Morecambe Bay area, at the mouths of the Leven, Kent, and Lune estuaries, northern England is a large overwintering population of knot (*C. canutus*). The knot breeds in the summer in the tundra of Greenland and NE Canada, where it feeds almost entirely on insects. In winter it moves south to the estuaries and coasts of Europe and North Africa and about 16% of the total population spends the winter in Morecambe Bay. The 70,000 birds at Morecambe Bay feed predominantly on *M. balthica*, supplemented by *M. edulis* and *H. ulvae*. The choice of *Macoma* appears to be dictated by its availability, as the knot feed predominantly on the lower half of the intertidal areas and thereby feed mainly on the *Macoma*, which are generally lower down the beach than *Hydrobia*.

The preferred diet of Redshank (*T. totanus*) at many estuarine sites in the British Isles is the amphipod *C. volutator*, and the feeding rate of Redshank depends mainly on the density of that prey in mud. Where the density of *Corophium* is low, they take the polychaete worms *N. diversicolor* and *N. hombergi* instead. The densities of both Redshank and curlew (*N. arquata*) on the Orwell, Stour, Colne, Blackwater, Crouch, and Roach estuaries in SE England are closely correlated with densities of their main prey (*C. volutator* and *N. diversicolor*). The distributions observed are due either to the birds responding behaviorally to the density of the prey in different estuaries and dispersing themselves in relation to it, or to the birds dispersing themselves in different estuaries but subsequently dying of starvation disproportionally in estuaries where food is scarce.

The relationship between bird usage of an estuary and prey density has been developed for five species of Wader on the Forth estuary, Scotland. Here there is a significant association

between feeding hours (i.e. number of birds and hours spent feeding per tide) and numbers per kilometer squared, and the density of at least one of their main prey. The significant associations, which explain up to 96% of the observed variation in bird distribution were found for Oystercatcher with mussels (*M. edulis*), for curlew, Redshank and Dunlin with *N. diversicolor*, and for knot with *Cerastoderma* (=*Cardium*) *edule*. Additionally, the distribution of bar-tailed godwit, knot, and Dunlin was related to the area of the intertidal mudflat, and Redshank to the exposure sequence (length of time exposed by low tide). The results of this study of bird distribution in relation to prey density are summarized in Table 5.6.

The steady fall in the numbers of Dunlin, *Calidris alpina* spending the winter in Britain may be caused by the spread of *Spartina* over the mudflats where the birds feed. It has been shown that the decline in the number of Dunlin varies from estuary to estuary. In estuaries where *Spartina* has not spread over the mudflats, the numbers have not changed, and the largest fall in numbers has been where the grass has spread most rapidly.

Studies on the Tees estuary, NE England suggest that up to about 90% of the macrofaunal biomass is removed by seven species of wader and duck. Most of the bird populations feed at the Tees during the winter time when the invertebrate prey is not undergoing growth, so it is therefore reasonable to compare the birds' feeding requirements with the biomass of the invertebrates. The birds select different prey items (see Table 5.7) to some extent, which enables them to live alongside each other. In the Sheepscot estuary, Maine, gulls (three *Larus* species) have been estimated to remove 6.8% of the worm *N. virens* population biomass, whereas in the Humber estuary, United Kingdom, birds consume only 2–3% of benthic production. Clearly estimates of bird predation vary, but a value of approximately 20% of benthic production being consumed by birds can be considered as an average value.

Table 5.6 Correlations for wader feeding-hours and wader feeding densities in kilometer squared in relation to invertebrate prey densities and site characteristics in the Forth estuary

	Oystercatcher	Curlew	Bar-tailed Godwit	Redshank	Knot	Dunlin
Macoma	X 0	X	X	X0	X	X
Cardium	X 0	X		X	X	X
Mytilus	X 0				X	
Hydrobia			X	X0	X0	X
Corophium				X		X
Nereis		X0	X	X0		X0
Nephtys			X			
Oligochaete				X		
Area			0		0	0
Inaccessibility Index					0	
Coverage sequence				0	0	

Note: Zero indicates a significant association in multiple regression analysis, and X indicates items found to occur frequently in the diet in the analysis of guts.

Source: Bryant (1998).

Table 5.7 Importance of different invertebrates in the diet of shorebirds feeding on seal sands, Tees estuary, England

Shorebird	Shelduck	Grey plover	Curlew	Bar-tailed godwit	Redshank	Knot	Dunlin	Energy content (invert. kcal x 10⁶)
	T. tadorna	*P. squatarola*	*N. arquata*	*L. lapponica*	*T. totanus*	*C. canutus*	*C. alpina*	
Invertebrate prey								
H. ulvae								
1 + year class	*	**	—	—	*	**	—	9.0
0–1 year class	*	—	—	—	—	*	*	3.0
N. diversicolor								
1 + year Class	—	**	**	**	*	*	*	9.4
0–1 year class	—	*	—	—	—	—	—	4.0
M. balthica								
All classes	—	—	—	*	*	*	—	0.6
Small polychaetes	**	—	—	—	*	—	**	85.5
Small Oligochaete	**	—	—	—	*	*	**	222.0
Total energy content of inverts.								333.5
Food requirements of birds (kcal × 10⁶)	73	2	23	9	17	93	68	
Total food requirements of birds (kcal × 10⁶)						285.0		

Note: *Indicates regular food items; **Indicates the chief food providing daily biomass; —indicates food item not taken at all or only occasionally. The total energy content of the invertebrates and the total food requirements of the birds are expressed as millions of kilocaloroie.

Source: Evans *et al.* (1979).

Birds, which feed intertidally in estuaries during daylight hours, may find conditions particularly severe during the wintertime at high latitudes. In a study of waders in the Wash, England, it was shown that, during the autumn, feeding conditions for the birds were adequate and over 30% of the available daylight hours were spent in non-feeding activities such as resting. During the short daylight days of the winter, however, they had to spend over 95% of the available hours feeding and even that may have been insufficient to sustain many individuals who apparently died due to food shortage. During the winter the Redshank on the Ythan estuary, Scotland, are only able to obtain 50% of their food from the estuary in daylight, and have to collect the balance at night or from the surrounding land.

Eider ducks (*S. m. mollissima*) feed by dabbling in the shallow water at the edge of the ebbing tide, or by diving into deeper water. The mussels they pick up are swallowed whole and crushed by the birds' large muscular gizzards. Such a feeding method precludes the intake of mussels larger than about 40 mm, and mussels less than 5 mm are rarely taken.

The gastropod snail, *H. ulvae*, is a major food source for wintering Shelduck (*T. tadorna*), and it is probable that most Shelduck concentrations, at least in Britain, occur where *Hydrobia* is abundant and available. The feeding activity of Shelduck can be classified into five methods, as shown in Fig. 5.6. These methods are:

1. Surface digging for *Hydrobia* buried in mud.
2. Scything action on exposed mud with a moist surface.
3. Dabbling and scything in shallow water.
4. Head dipping in 10–25-cm deep water.
5. Upending in 25–40-cm deep water.

Hydrobia displays a clear tidal rhythm, with the majority of individuals burrowing into the mud surface during low tide. As the tide reimmerses the mud, the snails reemerge onto the mud surface to commence their grazing behavior. By adopting the head dipping and upending modes of feeding, the Shelduck are able to catch the emerging snails, and to extend their time of feeding. Even at high tide many Shelduck manage to continue feeding, especially by upending. The scything action of the Shelduck is an effective means of sifting out *Hydrobia* and other surface-dwelling

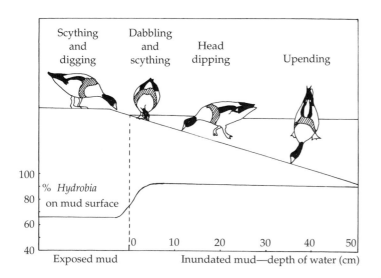

Figure 5.6 Shelduck feeding methods, showing the relationship between the different methods used and the water depth (cm), and the proportion of the *H. ulvae* population at the mud surface. (After Bryant and Leng 1976.)

animals such as Oligochaete from the mud surface, but its value at particular locations may depend on the moistness of the mud at low tide.

5.5 The impact of the secondary consumers

The secondary consumers as a group often efficiently utilize the food that is available to them, and also partition it between themselves in response to the pressures of ecological competition. As an example of this partitioning of resources three bird species exploit the same mussel beds of the Ythan estuary with herring gulls taking the 0-group mussels, eider the I-group, and Oystercatchers the II+-group. Of the summer production of 1300 kcal m^{-2} by the mussels, 112 kcal are consumed by the herring gulls, 275 kcal are consumed by the eider ducks and 93 kcal are consumed by the Oystercatchers. Man also removes 240-kcal m^{-2}, and the remaining 580 kcal m^{-2} is utilized by the mussels to support their metabolism in winter. Thus it may be seen in this population that the requirements of the predators and the metabolic requirements of the mussels are all provided by the production of the mussels.

We have already seen, however, that in the mudflats of the Ythan estuary as a whole the flounders and gobiid fishes consume three times the total amount consumed by the four major bird consumers. In various studies it appears that the fish and invertebrates are more important predators in terms of total consumption than the more conspicuous birds (Table 5.8). In some estuaries not all of the total primary consumer production is utilized by the secondary consumers. This apparent failure to consume all the available production may be due to several reasons. First, the seasonal movements of the predators due to migration in and out of the estuaries, or across continents, means that they are not always present to utilize the food available. Second, the limitations imposed by tidal

Table 5.8 Productivity of selected secondary consumers

Secondary consumer	Productivity (gC m^{-2} year^{-1})	Location
Estuarine birds	0.06	Dublin Bay, Ireland
Killifish (F. heteroclitus)	17.3	Tidal creek, United States
Flounder (P. flesus)	0.5	Ythan estuary, Scotland
Cape silverside (Atherina beviceps)	1.7	Bot estuary, South Africa
Clipfish (Clinus spatulatus)	0.05	Bot estuary, South Africa

Source: Wilson (2002).

and weather conditions may limit the time that can be spent feeding. Third, much of the production of the primary consumers will be utilized after death by the decomposer organisms, and through decomposition the organic matter becomes available again to the detritus feeders.

The clear impact of predators of the intertidal macrofauna of muddy areas can be demonstrated by placing cages on the mud, which exclude the predators and then observing any changes within the cages. In predator exclusion experiments in the German parts of the Wadden Sea, off the Island of Sylt, Reise (1985) has shown a remarkable (up to 23 times) increase of abundance of macrofaunal animals within exclusion cages as compared to control sites without cages. The muddy sediments of this area are inhabited principally by the small annelid worms, *Tubificoides benedeni*, *Heteromastus filiformis* and *Pygospio elegans*, with the main predators in the area the shore crab *C. maenas*, the shrimp *C. crangon*, and gobiid fish *Pomatoschistus microps*. Within cages of 1-mm mesh only a slight increase in the abundance of macrofauna was noted in the March to June period, but in the June to October period of study the numbers within the cage increased dramatically, while the numbers outside the cage decreased substantially. The bivalve spat within the cage grew substantially in this period, whereas extensive mortality occurred outside the cage. At the end of the study period the number of macrofauna within the cage were 86,475 m^{-2}, with the numbers outside 3750 m^{-2}. From a comparison of different mesh sizes, Reise was further able to show that in this part of the Wadden Sea the predation of shore crabs, shrimps, and gobies is responsible for major changes in the intertidal macrofauna, and that the impact of birds and larger fish is negligible. The results of such cage studies should always be treated with some caution, as the presence of the cage may alter the environment.

The numbers of birds that forage on the Dutch part of the Wadden Sea are at a maximum in autumn through to early spring and then low in late spring and summer. The numbers of birds clearly fluctuates independently of food supply, as the biomass of the prey species is at a peak in July and lowest in February. The factors, which control the season of peak numbers of birds must be sought elsewhere and relate to breeding migrations and the availability of food in other ecosystems. The numbers of fish and invertebrate predators, especially plaice (*P. platessa*) and shore crabs (*C. maenas*) in the Wadden Sea are closely linked to the maximum availability of their food species. As the biomass of prey species reaches a peak in June through to September so do the numbers of plaice and shore crabs.

In general, within the temperate mudflat-dominated estuaries of Europe there is a seasonal variation in the biomass of potential prey organisms with a maximum during the summer months. The migration of fish and invertebrate predators into the estuaries is often synchronized with this peak in the availability of the prey, and when calculated on an annual basis the fish and invertebrates are often clearly the main secondary consumers. Many estuaries particularly in Britain and The Netherlands see peak numbers of birds in the winter months, as the waders and wildfowl migrate there from eastern and northern Europe. Their areas of summer residence show an even greater seasonal variation in food availability with a rich abundance in the summer and virtual absence of food in the winter. The birds are thus driven to the estuaries of milder regions, where even though food may not be at its maximal biomass, it is at least available. The availability of prey species throughout the year is ultimately due to the dependence of the primary consumer on detritus as their main food source since this, almost uniquely amongst primary food sources, is available throughout the year, albeit at a somewhat reduced level in the wintertime.

5.6 Role of the secondary consumers in the estuarine ecosystem

We have seen that the primary consumers of estuarine intertidal areas are heavily exploited by the secondary consumers that visit the area. These visitors are primarily young fish, little crabs, and shrimps. The adults often stay in deeper water, while their larvae or juveniles come close inshore, feeding first on the meiofauna, and later on the macrofauna. Biomass is exported from the tidal flats in terms of

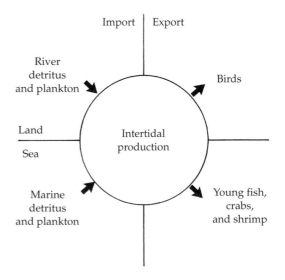

Figure 5.7 The tidal flat turntable for the flow of organic matter between the land and the sea. (After Reise 1985.)

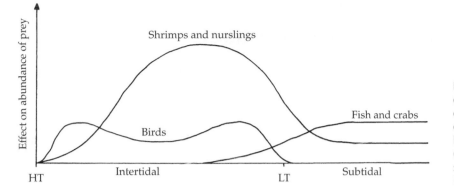

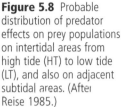

Figure 5.8 Probable distribution of predator effects on prey populations on intertidal areas from high tide (HT) to low tide (LT), and also on adjacent subtidal areas. (After Reise 1985.)

nurslings that have grown to adult size and return to the sea to grow and breed, and also to a generally lesser extent by birds, which feed on the same areas at low tide, and then fly off to the land at high tide. Reise (1985) in describing this situation uses the term "an ecological turntable" (see Fig. 5.7). Energy is received by intertidal flats from land and the sea as imported nutrients and detritus, and exported from the same area back to the land and sea by migrating birds, juvenile fish, and as dissolved nutrients. The various predator effects on the prey populations are summarized in a speculative manner in Fig. 5.8. The estuarine ecosystem is as we have already seen heavily subsidized by the adjacent ecosystems, but through the activities of the secondary consumers at least some of this energy is returned to support the marine and terrestrial ecosystems.

Sherman and Duda (1999) have recently reviewed the ecosystem approach to the global assessment and management of coastal waters, and show how worldwide the ecosystem approach is developing as a means of linking science—based assessments to the management of these resources. They warn, however, that management cannot wait for science to catch up with a full understanding of ecosystem structure and function. To be healthy and sustainable an ecosystem must maintain its metabolic activity level, must maintain its internal structure and organization, and must be resistant to external stress over time and space scales relevant to the ecosystem. The ecosystem can thus be examined on the basis of its productivity, its fish and fisheries, its "health" and its socioeconomic values.

On an economic basis the estuarine ecosystem provides many valuable services, which are often not traded in the stock markets of the world, but which are essential and valuable services. These include the services provided by estuaries and coastal wetlands as nurseries for fisheries, as natural pollution filters, and as storm buffers. Even though the estuaries are the recipients, from rivers and the land, of most of the pollution that subsequently enters the sea, by their ecosystem activity they manage to reduce the amount entering the sea, while at the same time providing essential feeding grounds for juvenile fish and migrating birds. Estuaries are a small proportion of the sea as measured by surface area or water volume, but it is clear that the ecosystem functioning of the estuaries that they provide an essential service to the ecosystem of the sea, far out of all proportion to their apparent size.

CHAPTER 6

Estuarine uses and users

6.1 Introduction

Mankind has long used sheltered estuarine areas such that now there are many uses of them—for providing fish and shellfish, aggregates for building such as sand and gravel, and water for abstraction. He has claimed land from the wetlands for agriculture and on which to build, and uses the area for recreation, and supporting infrastructure such as bridges, barrages, and harbors. They are the sites for industry and urban areas and so receive the discharge of wastes. They are major navigation waterways and thus there is the need for channels to be maintained by dredging. Each of these activities is considered to be legitimate in that they are often licensed, the users requiring official permission. Given those many uses, estuaries have many users—fishermen, industrialists, shipping, aquaculturists, and farmers. In addition, we can regard another user as wildlife, especially the ones that rely on these areas such as overwintering birds and marine fishes, which migrate into or through the estuary.

The human effects on estuaries can be separated into two major categories: (1) the input of materials in the wetlands and water column and (2) the removal of some part of the available resources. These inputs may be small or large particles, soluble materials that may sooner or later adsorb to other particles, and large structures such as bridges and harbors. The available resource removed by Man includes space, in which wetland area is lost through land-claim or, taking the concept further, the removal of a healthy water column by the addition of polluting materials. There

may be materials removed such as the bed of the estuary, by dredging to clear channels for navigation, and the winning of sand and gravel for building, or of sand for beach nourishment. Salt may be removed from the water column, by desalination plants for providing water for irrigation and the water may be abstracted by power stations for cooling. Biological materials may also be removed, for example, sea grasses, such as *Ruppia* or *Zostera* for fertilizer, and fish and shellfish for food.

Many of the problems encountered in estuaries are the direct result of the populations and infrastructure, which they support. Over one-third of the population of the United States lives and works close to estuaries, and seven of the ten largest cities of the world are situated adjacent to estuaries (London, New York, Tokyo, Shanghai, Buenos Aires, Osaka, and Los Angeles). In Britain, most of the main cities border estuaries and those estuaries also have highly developed and modified catchments, which drain agricultural, industrial, and urban areas. For example, the Humber Estuary's catchment drains one-fifth of England; areas previously dominated by the steel working, coal mining, and textile industries. The estuary has thus received waste for many decades.

Cities have developed on estuaries because of their role as natural transportation centers, providing sheltered, if sometimes shallow, harbors, and linking river and sea traffic. To supply the needs of commerce and navigation, quays and wharves have been built along the banks of estuaries where ships can tie up to load and unload. As ships have increased in size, so the pressures on the estuary have increased. First, it may be necessary to dredge the estuary to ensure a minimum depth for navigation. Second, as ships further increase in size, it may be necessary to extend the size of harbors or piers to accommodate larger vessels. Eventually, upper estuarine navigation routes are abandoned because of the size of modern supertankers, which cannot enter the old estuarial harbors, and new terminals have to be built in the lower reaches of the estuary. This process of a historical movement of estuarial harbors seawards has occurred in many of the world's estuaries, for example, on the Thames estuary from London down to Tilbury and Felixstowe, on the Clyde estuary from Glasgow down to Greenock and Hunterston, and on the Chesapeake estuarine system from Washington DC down to Norfolk. The abandoned harbors and docks may then be reused to provide up-market housing for city dwellers, complete with a view of their local estuary.

The use of estuaries for navigation and harbors has developed with industrial growth along their banks. Large-scale industry such as steel works, oil refineries, or chemical works need a combination of flat land and good transportation, as is usually found close to an estuary. Since the supply of flat land is finite, the increased demands of industry are often met by the "reclamation" (or more appropriately "land-claim" as the area is being taken for the first time) of estuarine mudflat. In extreme cases, such as the Tees estuary, United Kingdom, 90% of the former intertidal habitat has been claimed for land to provide flat land for industrial development (Fig. 6.1). Elsewhere, claiming an estuarial salt marsh may gradually increase farmland. Many of the harbors constructed in estuaries involve large-scale land-claim, and several "land-fill" sites for the disposal of domestic refuse involve claiming estuarine intertidal areas. To prevent against flood damage whole estuaries may be shut off from the sea, as has happened in the Netherlands,

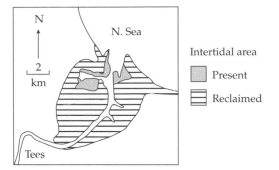

Figure 6.1 Land-claim of intertidal areas of the Tees estuary, NE England for industrial purposes over the last 60 years. (After Carter 1988.)

for example, in closing the estuarine Zuider Zee, to become the freshwater Ijselmeer, or in the large-scale Delta scheme in which the Grevelingen estuary has now become a brackish lake.

This chapter shows that all of these features are interlinked, especially as many users and uses occur at the same time. Many of the activities will change the structure of the estuary and add materials into the system. As shown throughout this book, all parts of the functioning of estuaries are linked. For example, any disruption to the estuarine hydrographic patterns can have a knock-on effect on the substratum and estuarine shape (the bathymetry and topography) and then to the structure and functioning of the biological system. Any changes to the system caused by man's activities thus require us to consider whether Man should respond and manage the system. The recently adopted "DPSIR" framework is useful in setting this in context (Box 6.1).

This chapter first considers the uses made of estuaries, and then examines the various responses of estuarine organisms to each usage. The chapter considers many forms of pollution, both separately but also, because they rarely act in isolation, in combination as well as other types of stress on the system resulting from Man's uses. There are well-defined categories of change due to

---

Box 6.1 The DPSIR approach

• *Driving forces*: the Drivers, the human activities responsible—this refers to "the big aspects," for example, the need for food such as fish and shellfish, the need for moving materials, the need for building materials, the use of space, the need to dispose of waste;
• *Pressures*: the causes of problems derived from the Drivers; each Driver will create many pressures, for example, emissions of waste, seabed damage through fishing, oil spills from tanker movements;
• *State*: the change in the environment, the background status, structure and functioning, environmental variables and physical, chemical, and biological features of an area or system;
• *Impact*: the changes to the human system, the Man-induced modification to the physics and

chemistry of the system, the biological responses at any level of biological organization (cell, individual, population, community, ecosystem), at any spatial scale (i.e. extent) or temporal scale (i.e. duration);
• *Response*: the "human" response to the causes and effects, the derivation and measurement of policy options, the need to ensure that the six tenets (environmentally sustainable, economically viable, technologically feasible, legislatively permissible, administratively achievable, socially desirable) are met for a sustainable management policy, the use of BPEO (Best Practicable Environmental Option), EIA (Environmental Impact Assessment), BAT (Best Available Technologies).

---

human activities in the estuarine ecosystem:

• the presence of xenobiotics (i.e. materials foreign to the natural system) and toxins
• physical additions
• energy change
• physical structures
• the overstimulation of biota
• the input of nonindigenous organisms
• and the production of a mutagenic response.

Similarly, there are categories of pollutants:

• trace metals
• synthetic organic compounds
• hydrocarbons
• radioactivity
• inert (physical) materials
• nutrients
• organic matter
• energy
• and alien organisms.

The changes in estuaries due to the above stresses can be regarded in the same way as any body that is functioning unusually or at a suboptimal level. In this way, we can think of estuaries subjected to stresses caused by

man's actions as having a pathology and so we can define and study that change as "attributes for the diagnosis of ecosystem pathology" (Harding 1992; Elliott and Hemingway 2002). The predominant changes to the system relate to:

• nutrients and productivity
• the presence of abiotic zones
• changes to species diversity
• changes to the size and biomass distribution of organisms
• disease prevalence
• the stability or instability of the overall biotic composition
• and the bioaccumulation of contaminants.

6.2 Estuarine pollution

Before discussing any topic, it is always necessary to define the terms being used. This is especially the case when we are discussing pollution and Man's effects on the estuarine system. Pollution is usually thought of in terms of chemicals, hence many of the definitions relate to the inputs of chemicals such as heavy metals, oils, radionucleides, nutrients and this has been the case since environmental awareness first increased. We should now, however, also include biological pollution, such as the input of organisms, and physical pollution, such as the input of biologically inert materials such as sediment, concrete, and structures, which will also have an adverse effect on the natural system. Box 6.2 gives the relevant definitions for chemical and biological pollution.

Because of the industrialized, agricultural, and urbanized nature of estuaries and their catchments, estuaries worldwide sooner or later receive much of the waste discharged by mankind into aquatic environments. Within the sea, almost all pollution is concentrated into estuaries and near shore coastal zones and so when environmentalists or politicians talk about pollution in the sea, they are often really discussing pollution in estuaries. The effects of pollutants may vary according to the chemical and physical state of the material being discharged into the estuary and any transformation after discharge, which is dependent on the chemical and physical characteristics of the receiving waters. The effects of pollutants on the estuarine ecosystem will also vary both seasonally and temporally, as well as being related to environmental factors, such as water circulation and salinity. The effect of a particular pollutant may vary according to the part of an estuary, which receives it, for example, sewage discharged into the head of an estuary in summer may consume all the oxygen, and bring about major changes in the ecosystem, whereas the same quantity of sewage discharged into the mouth of an estuary in winter may have so little effect as to be virtually unmeasurable. This ability to absorb an amount of a contaminant added to the system is regarded as the "assimilative capacity." The assimilative capacity of a small stream is easily exceeded by a discharge of sewage and so Man has successively moved the main waste streams to larger bodies of water—first large rivers, then estuaries, and now open seas.

The responses of estuarine organisms to pollution range from the acute to the minimal. At the highest level of pollution, the responses of the animals and plants are easily recognized, since the results are acute and may be lethal to all forms of life. At a lower level of pollution, the sensitive fauna is eliminated, but tolerant species may thrive and become more abundant. The estuarine ecosystem thus shows distortions of conditions and organisms. Measurement of these distortions is the basis of the biological monitoring of pollutant effects. At the lowest level of pollution, only subtle changes in the physiology and biochemistry of the organisms may occur. Such subtle changes may be crucial in the long term, but can be difficult to detect.

All species are tolerant of a certain amount of environmental variation and, as shown in earlier chapters, estuaries are variable and estuarine organisms are thus more tolerant to wider environmental variation than other systems. The tolerance to pollution will certainly vary from species to species, and will often vary between individuals of the same species. Given that estuarine species are already particularly tolerant to environmental variation compared to marine or freshwater species, then the capacity of the estuarine ecosystem to accept pollutants, which enhance natural variation, such as organic matter, is relatively great. Because estuaries are such highly variable and thus perhaps resilient systems, it becomes more difficult to detect man-induced (anthropogenic) stress in the biota separately from natural stress and from natural variability.

"Hot-spots" of acute pollution are usually readily detectable, and often incur public concern, but long term, more subtle, "chronic" pollution is often sanctioned by governments, and is less likely to feature in newspaper headlines, despite having profound effects on the estuarine ecosystem. For example, an oil spill at the mouth of an estuary will command immediate attention because of the pollution risk, whereas an industrial discharge containing only a small percentage of oil will rarely

Box 6.2 Definitions of terms used in chemical pollution and their translation for biological pollution (modified from Elliott 2003)

Term	Chemical-based definition	Biological pollution
Pollutant	A substance introduced into the natural environment as a result of man's activities and in quantities sufficient to produce undesirable effects.	The input and effects of micro- and macro-organisms to produce adverse effects.
Pollution	(i) is a change in the natural system as a result of man's activities; (ii) has occurred if it reduces the individual's/population's/species'/community's fitness to survive. The introduction by man, directly or indirectly, of substances or energy into the environment resulting in such deleterious effects as to harm living resources, hazards to human health, hindrance to marine activities including fishing, impairment of quality for use of seawater, and reduction of amenities (GESAMP).	The effects of introduced, invasive species sufficient to disturb an individual (e.g. internal biological pollution by parasites or pathogens), a population (by genetical change) or a community (by increasing or decreasing the species complement); including the production of adverse economic consequences.
Contamination	An increase in the level of a compound/element ("pollutant") (as the result of Man's activities) in an organism or system which not necessarily results in a change to the functioning of that system or organism.	The introduction of species without noticeable effects (e.g. microbes killed in seawater; species occupying available and vacant niches).
Responses to pollution	(i) lethal—organisms are killed thus resulting in community change; (ii) sub-lethal—effects which may occur before the concentration of toxic substances reaches lethal levels.	(i) the introduction of pests which increase predation/mortality; (ii) an increase in an immunological response in an individual due to exposure to microbes.
Stress	The cumulative quantifiable result of adverse environmental conditions or factors as an alteration in the state of an individual (or population or community) which renders it less fit for survival.	A reduction in health due to pathogens and parasites and the loss of genetic fitness due to escapees. The alteration in community structure with single invasive species at low densities may be difficult to detect.
Episodic pollution	Major but often short-lived (temporary) discharges, e.g. shipping accidents.	Discharges of organisms via ballast-water may be regarded as episodic although there may be an insufficient influx to establish a population.
Chronic pollution	Diffuse, low-level inputs which cannot be traced to a particular incident, e.g. from rivers, atmosphere; may contribute 90% of total inputs; often the cause of effects which are sub-lethal and difficult to detect.	Regular, continuous inputs of organisms as escapees or the dispersive, reproductive stages of non-native cultured organisms that have adapted successfully and can reproduce; gradual movement of species through climate change.
Acute pollution	Concentrated pollution, with an identifiable source and readily observed effects; often the cause of lethal effects.	Pathogenic micro-organisms discharged from waste-water outfalls.
'Aesthetic pollution' (sic)	Unpleasant material likely to cause visual or olfactory offence but which (usually) causes little biological harm.	The aesthetic aspects of changes to natural faunal and floral communities, i.e. a reduction in 'naturalness'.
Bioaccumulation	An increase with time in the content (or body burden) and/or concentration of a contaminant within an organism.	The uptake and accumulation (culture) of pathogens (and/or cysts) in filter-feeding bivalves may occur. The accumulation after successful reproduction of invasive species in a community.
Biomagnification	Increasing levels of a pollutant from the lower trophic strata to the higher trophic strata.	The successful establishment (and displacement of indigenous species) of an invasive population.
Direct input	A point source discharge, often the cause of acute pollution.	The release of pathogens from waste-water discharges, of genetically-modified organisms from culture, or parasites from transplanted shellfish.
Indirect (or diffuse) input	A widespread, low-level discharge often likely to result in chronic pollution.	The changes in distribution of species through man-induced global climate change.
Conservative wastes/inputs	Materials with slow degradation rates (long half-life), more likely to bioaccumulate, e.g. trace metals, halogenated hydrocarbons, and radioactivity.	The accumulation in the marine environment of successfully-colonising organisms, to the extent that their population becomes established and increases.
Degradable wastes/inputs	Materials more liable to be assimilated after being degraded through physical, biological and chemical processes in the short to medium term, e.g. sewage, hydrocarbons, pulp/paper wastes.	The demise and degradation of micro-organisms (e.g. faecal coliforms) in seawater after liberation from waste-waters. The unsuccessful colonisation of introduced species.

feature in the public domain, despite the fact that over long timescales, more oil could enter the estuary from the industrial discharge than is spilt from a stricken tanker.

6.3 Sewage pollution

Man has long used estuaries for the disposal of waste material and sewage is discharged into many estuaries. In many cases the raw untreated sewage is discharged, and in other cases the sewage is treated on land in sewage treatment works and only the liquid effluent produced is discharged into the estuary. The waste so discharged may, if there is only a little, become incorporated into the estuarine ecosystem as another source of detritus. The quantities discharged may, however, be so great as to cause major changes to the fauna and flora. The solid waste from sewage works, known as sewage sludge, has to be disposed of, and several countries including the United States have allowed the disposal of such material to sea or estuary. Prior to 1998, the United Kingdom and Ireland dumped sewage sludge into the sea but this was then banned in Europe under the European Commission's Urban Waste Water Treatment Directive and other international conventions.

Whether sewage works provide primary treatment (the removal of solids only) or secondary treatment (full biological treatment of liquid waste as well), considerable volumes of solid waste, sewage sludge are produced. The coastal area was long regarded as the most economical method of disposing of this sludge by transporting the waste by ship from the sewage-works to a dumping site at the mouth of the estuary, or further offshore. Two types of dumping site can be distinguished, either accumulation sites where the material accumulates on the bottom, or dispersal sites where currents disperse the material over a wide area, hence using the ability of the sea to disperse, degrade, and assimilate the waste. For example, a major sludge-dumping site is in New York Bight, where large quantities of waste from New York

City have been dumped near Ambrose Light. The bottom water oxygen concentrations are depressed at and around the dumpsite, and in spring and summer 1976 an area of 12,000 km^2 experienced oxygen concentrations less than 2 mg l^{-1}, with a smaller area becoming completely anoxic, resulting in the death of at least 143,000 tonnes of the clam *Spisula*. Reporting the situation in 1980–2, Steimle (1985) has shown that there is not an azoic "biological desert" in the dumping area, but the benthic fauna showed enhanced production of the stress-tolerant species, leading to an enhanced benthic biomass (127–344 g m^{-2} wet wt) with productive demersal fisheries.

Prior to the banning of seaborne sewage sludge disposal in 1998, the dumpsite for the sewage of the city of Glasgow was at Garroch Head, where 1.5 million tonnes of sewage sludge were dumped annually. This was a typical accumulating site, and the biomass and abundance of the fauna was very high at the center of the dump site (being 500 g m^{-2} and about 500,000 individuals m^{-2}) although the species diversity was low with the principal species being nematodes (*Pontonema* spp.), the Oligochaete *Tubificoides benedeni* and the polychaetes *Capitella capitata* and *Scololepis fuliginosa* (Pearson *et al.* 1986). The species diversity increased away from the dumping site, while the biomass and abundance declined (Fig. 6.2). These changes in species diversity, abundance, and biomass conform well to the general pattern described below, termed the Pearson–Rosenberg model. It was of note that this sludge, from a highly industrialized city, also contained high levels of persistent pollutants such as heavy metals that then were retained in the sediments and the fauna. Whereas many of these pollutants were retained at the disposal site, some were exported in migrating fish feeding on the high levels of invertebrates at the site. Despite all of this, the site was regarded as a marine "land-fill site."

In contrast to the Garroch Head site, the sewage from the city of Edinburgh (up to 500,000 tonnes annually) prior to 1998 was

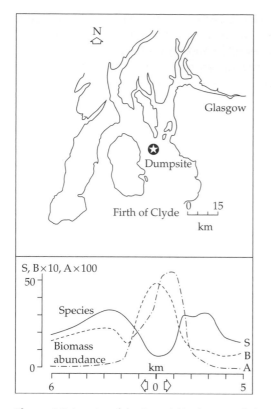

Figure 6.2 Location of the Garroch Head sewage sludge disposal site in the outer Clyde estuary Scotland, in upper diagram. The lower diagram shows the biomass (B, g wet wt m^{-2}), total abundance (A, m^{-2}) and species number (S) of the benthic infauna along an 11-km east/west transect across *the* disposal site. (After Pearson, Ansell and Robb 1986.)

dumped at two alternating sites, in summer near Bell Rock and in winter near St Abb's Head off the eastern Scottish coast. The 6-monthly disposal strategy at these dispersing sites was designed to allow recovery out of the respective dumping seasons. The dispersing nature of the sites was found not to produce evidence of any organic carbon accumulation in the sediments, and only slight changes in the benthic fauna. Hence, the sludge was sufficiently dispersed not to cause any significant alteration to the benthic environment. Similarly, again prior to 1998, 5 million tonnes of sewage from the London area were dumped annually on a dispersing site in the Thames

estuary, again with no major adverse environmental effects being noted.

In the 1990s in Europe, concern was expressed at the ethics and desirability of dumping sewage sludge at sea, since even with full dispersal a large amount of organic matter and nutrients have still been added to the ecosystem. Defenders of the marine disposal option considered it as a wise use of the environment, in that clean sewage sludge, that is, which is primarily domestic sewage sludge with little persistent pollutants, would be effectively degraded and assimilated in the sea. The alternative methods for disposal are incineration, drying, placing in landfill sites, spreading on farmland or spraying into forestry areas. Despite claims that the marine option was the best practical environmental option, these arguments were not accepted thus leading to the banning of sea-disposal within Europe in 1998.

6.4 Diffuse sources of nutrients

Although sources such as sewage, or mariculture waste do contribute some extra nutrients to estuaries, it has been widely shown that the major source of extra nitrogen is from rivers, due to drainage from the fertilizers used in agriculture, and from inland sewage works. It has been generally estimated that less than half of the fertilizer applied to agricultural lands becomes bound into the harvested crop, the remainder is lost through drainage into rivers and thus enters estuaries, and they also enter directly from fields adjacent to estuaries. Atmospheric deposition of nitrogen is the second most important source, after fertilizers. The supply of agricultural nutrients to estuaries does stimulate plant production, but not all of the nutrients are, or can be, utilized within the estuary, often because of either the short residence time or the low light (high turbidity) limiting conditions and the nutrients generally pass out from the estuary into the adjacent sea. Within Europe, there is concern that the North Sea is threatened by such excess nutrients that are entering it from the many estuaries bordering it. These nutrients are derived from inland sources, and carried from the rivers, through the estuaries into the sea. Without doubt the biggest source of such nutrients is the river Rhine.

The addition of extra organic matter (carbon) to estuaries and the sea is generally known as *organic enrichment*, and the increase of nutrients (nitrogen, phosphorus) from the same source is known as *hypernutrification*. Hypernutrification can lead to a set of adverse effects under the right environmental conditions and those adverse effects are known as *eutrophication*. Eutrophication may be natural or from mankind, that is, anthropogenic. In freshwaters it is the supply of phosphates that may cause eutrophication, but in estuaries and the sea it is usually the supply of nitrogen, which is the critical factor. De Jonge and Elliott (2001) and Elliott and de Jonge (2002) indicate the causes and consequences of nutrient enrichment as a set of "bottom-up causes" and "top-down responses." High inputs of nutrients require optimal conditions of a long residence time and good light conditions in order to be used for plant growth. In usually turbid estuaries, the nutrients may occur in elevated levels but do not generally lead to excessive algal growth because of light limitation. If there are good conditions for the use of the nutrients then the symptoms of eutrophication can occur. These are:

(1) increased macroalgal growth as macroalgal mats such as *Enteromorpha*, which may be floating in shallow brackish waters, as along the western Swedish coast, or on intertidal flats such as the Tees estuary, NE England;
(2) blooms of planktonic microalgae, which may be toxic, such as the red-tide algae producing Amnesic, Paralytic, or Diarrhoeic Shellfish Poisoning (ASP, PSP, DSP) such as *Alexandrium* and *Cingula*, or just a nuisance by producing large amounts of mucilage and foam such as *Phaeocystis*;
(3) decreased water transparency as the result of the blooms;

(4) oxygen depletion resulting from the dieback of the increased plant growth, which kills fish and benthos;
(5) organically enriched sediments following the dieback of the plant material and changes to the benthic community.

Rosenberg (1984, 1985) showed that eutrophication has been known to be occurring in the Baltic Sea for some time, as a consequence of the large nutrient inputs to the area coupled with the long residence time of this brackish sea. During the 1980s, the symptoms of eutrophication were also found in the Kattegat, the shallow sea between Denmark and Sweden, with marked changes in the bottom fauna observed, and mass mortality of bottom-dwelling animals such as *Nephrops norvegicus* (scampi, or Norway lobster). The demersal fisheries (for cod, etc.) declined, but the pelagic fisheries (for herring, etc.) increased.

The phytoplanktonic blooms may be toxic to other organisms. In 1987 such a "red tide" bloom of toxic algae forced the closure of 170 miles (250 km) of clam and oyster beds in North Carolina. Similarly, in the late 1990s, blooms of these organisms in west Scottish coast sea-lochs were thought to be associated with increased nutrients and organic matter from fish farming. Eutrophication has now been recorded in many estuaries of the world; as an example, the steady increase of phytoplankton in the Ems–Dollard estuary in response to eutrophication, measured here as total phosphates, is shown in Fig. 6.3.

On the intertidal mudflats of many estuaries, for example, the Ythan estuary in NE Scotland and the Tees estuary in NE England, the growth of extensive "algal mats" is regarded as a clear symptom of eutrophication. Algal mats of *Enteromorpha* or *Cladophora* smother the primary consumers in the benthos, and prevent oxygen reaching the sediments as well as increasing oxygen uptake by the degradation of the organic matter. This

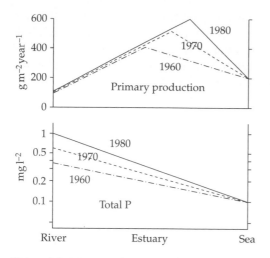

Figure 6.3 Primary production of phytoplankton, and total phosphorus (P) in the Ems–Dollard estuary in 1960, 1970, and 1980. Note the eutrophication related to increased supply of P from river and inland sources. (After Postma 1985.)

produces anoxic conditions, such as hydrogen sulfide production, which in turn kills animals such as the bivalves *Cerastoderma* or *Mya*, but increases grazers such as *Hydrobia*. The secondary consumers such as the wading birds and fishes are prevented from feeding on the mudflats, thus the whole system has been adversely affected.

The consequences of organic enrichment of estuaries are thus shown to be variable, and many types of assessment are required to interpret their effects. Understanding the functioning of the estuarine ecosystem lies at the heart of our comprehension of organic pollution. Organic enrichment in small amounts may produce "acceptable" changes in estuarine systems, whereas large amounts may produce "catastrophic" changes resulting in mass mortality. As Van Impe (1985) has clearly shown, a long-term increase in the numbers of estuarine birds does not always appear to be an indication of an improvement of the environmental quality of an estuary, it may in fact indicate an increase in estuarine pollution. In the Scheldt estuary an

incontestable ecological deterioration, due to several sources of mainly organic pollution, has led to a greater food supply for intertidal birds. The birds have responded by visiting the area in increased numbers, and by staying longer. In contrast to other forms of pollution, the effects of organic pollution are usually reversible, so the estuarine ecosystem with its in-built flexibility can accommodate a moderate amount of organic pollution, as well as recover fairly rapidly from severe organic pollution.

In a budget of organic inputs to the Humber estuary, Eastern England, Elliott *et al.* (2002) estimated that direct inputs from anthropogenic sources accounted for 50% of the total (Table 6.1) and that the amount from estuarine macrophytes had decreased greatly due to the reclamation of large areas of estuarine wetland. It was argued that the functioning of the estuary had remained the same as historically because natural inputs fueling the system had been replaced by anthropogenic ones. Of course, this becomes a matter of concern if those anthropogenic sources are subsequently removed through increased treatment such that there is less organic matter supporting the detritus-based system.

Table 6.1 The organic budget of the Humber Estuary, England, before (<1700s) and after (2001) land-claim

Inputs	tC year^{-1} (<1700)	tC year^{-1} (2001)
Industry	0	1,856
Sewage, STW	0	4,631
Sewage, CSO	0	4,329
Natural vegetation [*]	22,900	12,205
Diffuse sources via freshwater (FW-sewage)	4,924	4,924
Total	**27,824**	**27,945**

Note: Historically, 2300 ha of intertidal have been lost (Murby 2000), it is assumed that 50% of this was salt marsh and 50% mudflat, then converted from organic matter.

Source: Boyes and Elliott, unpublished.

6.5 Organic enrichment

A small amount of organic matter, well dispersed, can be readily utilized within the estuarine ecosystem to enhance the levels of biological production. Problems arise, however, if a large volume of organic effluent is discharged into an estuary with a low flushing rate, or if too much effluent is discharged at a single point, that is, the assimilative capacity of the receiving area is exceeded. When excess quantities of organic matter are present, then bacteria and other microorganisms, which utilize the organic matter, will consume all the available oxygen in the water. Such problems may be particularly severe in summertime because of reduced river flow, shallow water conditions, coupled with temperature increases, which lead to enhanced bacterial activity, and thus accelerate the depletion of the oxygen content of the water. The shallow nature of the upper estuarine areas then exacerbates this problem. Figure 6.4 shows the water column with dissolved oxygen levels in the upper Forth estuary and in particular illustrates both a seasonal (i.e. summer) and spatial (i.e. upper estuarine) dissolved oxygen sag or minimum.

The development of low oxygen, or even anoxic, conditions leads to the extinction of the normal macrofauna of the estuary, and its replacement by annelid worms, especially Oligochaetes. In areas of greatest organic enrichment even such worms may be excluded. The microbial degradation of organic waste will, apart from consuming the available oxygen, also lead to increased sulfide production. The oxidation of sulfide leads to further oxygen depletion. The effects of organic enrichment are most clearly seen by monitoring the fauna and measuring the redox potential (E_h) of the sediment at successive distances away from a known organic discharge. The redox potential may be determined with special platinum electrodes inserted into the sediment,

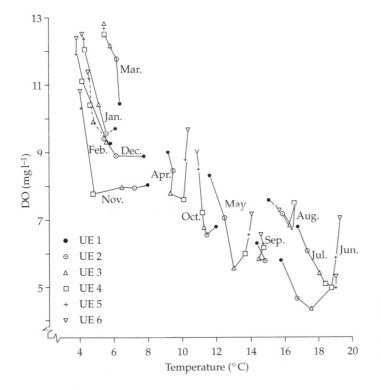

Figure 6.4 Dissolved oxygen and temperature profiles of the Upper Forth Estuary, Scotland, showing monthly averages. (Forth River Purification Board, Edinburgh, 1988–90 data, UE1 in mid-estuary and UE6 at the tidal limit.)

with the resultant negative E_h values indicating reducing (anoxic) conditions, while oxygenated sediments have positive E_h values. Pearson and Stanley (1979) give full details of the measurement of E_h.

The results of the measurement of fauna and E_h at distances of up to 800 m from the discharge of organic effluent from an alginate (seaweed) factory are presented in Fig. 6.5. Close to the effluent were relatively few species, with low numbers and biomass, coupled with negative E_h values. At 100 m the abundance of animals (mainly small polychaetes) increased dramatically, but there are still few species. The biomass and E_h increased more steadily. Between 100 and 400 m the total numbers (abundance) of animals decreased, while the species diversity and E_h steadily increased. The biomass showed a dip at 200 m, but otherwise increased. Beyond 400 m the species diversity, abundance, and biomass all fell, while the E_h continued to rise. Such patterns of change in the species diversity (S), total abundance of animals (A), and total biomass (B), have been recorded in the vicinity of many organic discharges, and the form of the curves have become known as "S-A-B curves" or the Pearson–Rosenberg model (Pearson and Rosenberg 1978). Such S-A-B curves are now regarded as typical for the effects of organic enrichment, and can be interpreted as follows. Close to the effluent, conditions are anoxic, and no macrofauna can live. As one moves away from the effluent the sediment is colonized by "opportunistic" animals, mostly small annelids, which occur in enormous numbers. This zone, the "peak of the opportunists" has low species diversity and modest biomass. Beyond this is the "ecotone," where the more normal fauna replaces the opportunists, and as shown at 200 m in Fig. 6.6, a decrease in the biomass may occur. Beyond the ecotone, is the "transition zone" where the fauna shows enrichment due to the organic source, with maximal biomass and species diversity. Beyond this enriched zone the fauna declines to its background levels.

The change in species composition is shown pictorially in Fig. 6.6. Water movement and renewal is an essential factor affecting the impact of organic material in a particular area. The importance of water renewal is shown in Fig. 6.7. The importance of water renewal is such that a given organic discharge into the well-mixed mouth of an estuary may have little or no measurable effect, whereas the same discharge into the restricted confines of the head of the estuary may produce severe symptoms of organic enrichment.

The nature of estuarine and marine communities under both polluted/stressed and clean/unstressed conditions can be listed according to the changes in the primary community parameters of number of taxa, abundance, and biomass (Table 6.2). While these features differ to a small extent with different types of stressor, the general features hold for many stressors.

6.6 Overall system change

The Wadden Sea is the largest estuarine area in Europe, occupying about 10,000 km^2 and is the principal haunt of millions of shore birds as well as being a key nursery area for the

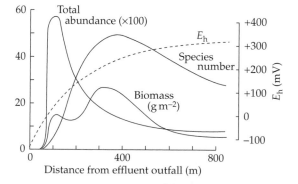

Figure 6.5 Species number, total abundance of animals, biomass, and E_h along a gradient of decreasing sedimentary organic content in Loch Creran, Argyll, Scotland, expressed as distance from the effluent of a seaweed processing plant. (After Pearson and Stanley 1979.)

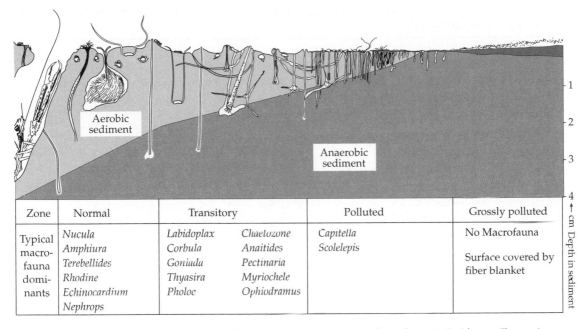

Zone	Normal	Transitory		Polluted	Grossly polluted
Typical macro-fauna domi-nants	*Nucula* *Amphiura* *Terebellides* *Rhodine* *Echinocardium* *Nephrops*	*Labidoplax* *Corbula* *Goniada* *Thyasira* *Pholoe*	*Chaetozone* *Anaitides* *Pectinaria* *Myriochele* *Ophiodramus*	*Capitella* *Scolelepis*	No Macrofauna Surface covered by fiber blanket

Figure 6.6 Diagram of the changes in fauna and sediment structure along a gradient of organic Enrichment. The species listed are typical macrofauna species for north European marine waters and equivalent species will be found elsewhere. (From Pearson and Rosenberg 1978.)

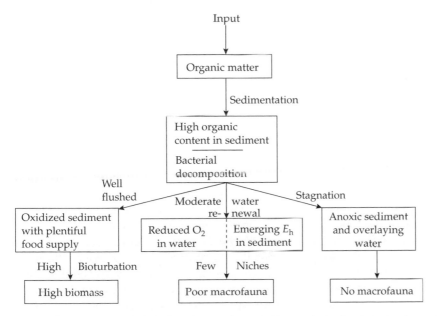

Figure 6.7 Diagram to show some pathways of organic input to the marine and estuarine environment, and its effects in relation to water renewal. (From Pearson and Rosenberg 1978.)

Table 6.2 Conceptual basis and assumptions inherent in macrobenthic impact studies. From Elliott (1994) modified and expanded from McManus and Pauly (1990)

A. Natural state

A natural macrobenthic assemblage either tends toward or is in an equilibrium state;

Under non-impacted conditions, there are well-defined correlation and relationships (which therefore may be modeled) between faunal and environmental (abiotic) variables;

In approaching the normal equilibrium state, the biomass becomes dominated by a few species characterized by low abundance but large individual size and weight;

Numerical dominance is of species with moderately small individuals, this produces among the species a more even distribution of abundance than biomass;

The species are predominantly K-selected strategists;

B. Moderate pollution

With moderate pollution (stress), the larger (biomass) dominants are eliminated, thus producing a greater similarity in evenness in terms of abundance and biomass;

Also, with moderate pollution, diversity may increase temporarily through the influx of transition species;

C. Severe pollution

Under severe pollution or disturbance, communities become numerically dominated by a few species with very small individuals;

Those small individuals are often of opportunist, pollution-tolerant species, which have r-selected strategies;

Under severe pollution, any large species that remain will contribute proportionally more to the total biomass relative to their abundance than will the numerical dominants;

Thus under severe pollution, the biomass may be more evenly distributed among species than is abundance;

However, under severe pollution, species with large individuals may be so rare as to be not taken with normal sampling;

The change in assemblage structure with increasing disturbance is predictable, follows the conceptual models and is amenable to modeling and significance testing;

D. Recovery

Opportunists are inherently poor competitors and may thus be out-competed by transition species and K-strategists if conditions improve;

McManus and Pauly (1990) also consider that under normal conditions:

The biomass-dominants will approach a state of equilibrium with available resources;

The smaller species are out of equilibrium with available resources;

The abundances of the smaller species are subject to more stochastically controlled variation than the larger species.

---

fishes of the North Sea (Fig. 6.8). Organic pollution of the Wadden Sea reflects the distribution of populations and industry, with untreated or minimally treated effluent from dairies and slaughterhouses in the Danish sector, effluent from the estuaries of the Elbe and Weser in the German sector, and from major Dutch cities. Viewing the Wadden Sea as a whole the organic effluents cause severe local problems of oxygen deficiency, but on a gross scale the area remains as probably the most important and productive estuarine ecosystem in Europe.

One estuary, which has recovered dramatically from decades of organic pollution, is the Thames estuary, which flows through London, United Kingdom. Due mainly to massive amounts of domestic sewage the state of the Thames declined steadily during the first half of the twentieth century, until by the 1950s it was completely anoxic over the middle reaches, inhabited only by Oligochaete worms, and mallard and swans, which fed on grain spillages. Conditions only began to improve in 1964 with the completion of a new sewage works at Crossness, and improved

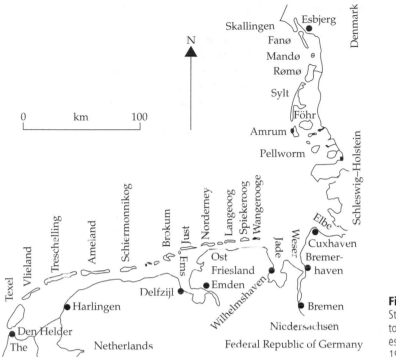

Figure 6.8 Map of the Wadden Sea. Stretching from Texel in The Netherlands to Esbjerg in Denmark this is the largest estuarine area in Europe. (After Essink 1978.)

further in 1974 due to another sewage works at Beckton, coupled with the closure of several old discharges. As a result of these improvements the condition of the estuary water has steadily improved, and by 1975 the lowest oxygen content was 25% of saturation. Since then the water quality has steadily improved, and as it has done so, fish and birds have returned to the estuary. Figure 6.9 records the number of fish species caught during the recovery period. In the period 1920–60 no fish were caught. The eel was the first to return, and by 1975, 36 species were found, including most dramatically the return of salmon, *Salmo salar*. The number of fish species has increased gradually with, by 2002, over 120 species being found. Similarly, a wide diversity of invertebrate fauna now occurs, although there are fewer Oligochaetes. As the Oligochaetes have reduced so have the numbers of tufted duck and pochard, which fed upon them. The numbers of wading birds and other ducks have increased dramatically in response to the return of a wide diversity of prey organisms.

6.7 Fisheries and aquaculture

6.7.1 Wild fisheries

Worldwide, the main estuarine fisheries are for shellfish, with the collection of some of the abundant natural populations of invertebrates, such as shrimps, crabs, oysters, cockles and mussels. Increasingly the catching of natural stocks is supplemented by mariculture, most notably for oysters and mussels. Bait digging by sport fishermen occurs widely in estuaries, and may involve considerable disruption of the intertidal fauna.

In addition to the shellfish industries are the vertebrate fisheries. While many of the fish, which enter estuaries are not commercially

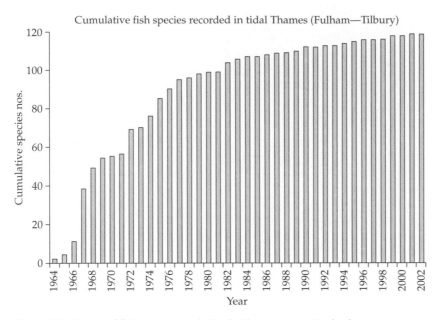

Figure 6.9 Number of fish species recorded in the Thames estuary, England. Drawn from data supplied by Dr. S. Colclough, Environment Agency, London.

exploited because they are the nursery stocks, or unwanted species, other species are heavily exploited. Salmon, sea trout, and eels all pass through estuaries on route from the sea to rivers, and many commercial fisheries exploit them. The Gulf of Mexico, stretching from Florida to Texas and Mexico, is a key fishery area for the United States, with 28% of their total fish landings coming from the brackish bays and lagoons of this area. The catch is mainly menhaden *(Brevoortia* spp.), striped mullet *(Mugil cephalus)*, croaker *(Micropogon undulatus)* plus Penaeid shrimps, blue crab, and oysters. All except the croaker are estuarine species, which in general spawn at sea. The larvae enter estuaries, grow there, and when fully mature return to the sea. Other rich estuarine fisheries are at the mouth of the Amazon, in Nigerian estuaries, and in Indian estuaries, such as the Ganges, where estuarine conditions extend 160 km upstream from the mouth.

As with all types of fishing, there are now major concerns on the ecosystem effects of the activities, these have been summarized in Fig. 6.10. For example, as well as the removal of target and nontarget fish and shellfish, fishing activity in estuaries can disturb or destroy benthos and thus affect their dependent fish and bird populations (Jennings and Kaiser 1998; Hall 1999; Blaber *et al.* 2000). The main effects of fishing are:

(1) to cause local extinctions, where there is a large-scale removal of a particular species, often as the target species, for example, the smelt, *Osmerus eperlanus*, which disappeared from the Forth Estuary, Scotland in the middle of the twentieth century mostly as the result of over-fishing;

(2) to affect the population viability, genetics, and maturity of the target organisms, for example, the removal of spawning stocks through over-fishing of pelagic and demersal populations;

(3) to affect nontarget organisms, both other fishes, such as the taking of juvenile herring during estuarine fishing for sprat, and other

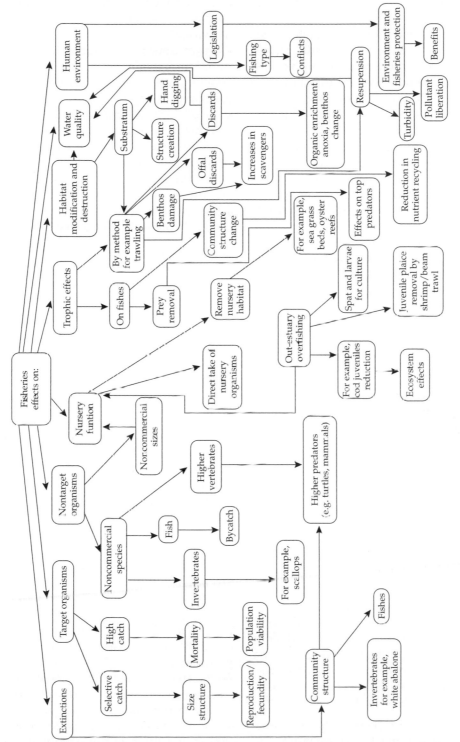

Figure 6.10 The effects of fisheries or estuarine ecosystems. (From Elliott and Hemingway 2002.)

species such as porpoises taken in estuarine salmon netting;

(4) to affect the nursery function, such as the removal of large numbers of juvenile plaice and dab during shrimp fishing, and by removing the mudflat habitat used by juveniles;

(5) to affect trophic interactions such as the removal of prey fishes, such as sandeels taken by seabirds and crustaceans, or the shrimps, which are central to the functioning of many estuarine food webs, and by changing food webs by increasing scavengers, especially benthic megafauna, following seabed damage by fishing gear and the production of detrital bycatch;

(6) by habitat modification and destruction, through land-claim and infrastructure creation (the building of ports and harbours) and by changes to substratum integrity by trawling;

(7) by water quality effects, which may occur through bed resuspension caused by trawling, the input of other introduced materials such as litter, and through organic enrichment produced by discards.

Many of the problems caused by marine fisheries occur in estuarine and wetland habitats although there are few case studies. Elliott *et al.* (1990) suggests an example of a marine fishery affecting estuarine fishes where the coastal and marine cod fisheries appeared to affect the nursery function of estuaries. In the Forth estuary, eastern Scotland, the decline of juvenile cod appeared to reflect the increasing North Sea catches of the larger, reproducing stocks. This feature was also shown in the Tyne estuary, NE England.

Each type of fishing will have an ecosystem effect and so, as an example, trawl fishing in temperate estuaries is considered here (Fig. 6.11). The summary (Conceptual Model) concentrates on the effects of trawling and its related activities (port and harbors, litter production and gear loss). This conceptual model summarizes the effects on the following aspects:

• noncommercial sizes (especially juveniles and thus the nursery function of estuaries)

• noncommercial species (the remainder of the assemblage and the effects of discards and individual damage)

• target species (through community and population changes at the ecological and genetic levels, and the effects on competing predators)

• the physical integrity of the system (through bed disturbance and damage to the benthos)

• contamination of the estuarine system (through discharges of soluble pollutants and large and small particulate materials, including litter and gear loss)

• the creation of infrastructure (habitat loss, especially of highly productive intertidal areas).

At present, there is no attempt to quantify the links within the conceptual model or to assess the geographical extent of the effects and so the model remains qualitative rather than quantitative.

6.7.2 Aquaculture (culture of finfish and shellfish)

Since the 1980s there has been the rapid growth of mariculture in many countries. In Scotland, Norway, and Canada the main expansion has been in the rearing of salmon in floating cages (Fig. 6.12). In assessing the effects of aquaculture, it is necessary to consider the areas used, the conditions required by the fish farmers or encountered in those areas, the effects on the water column, sediment and benthos, the impacts on local fishes and birds, the impacts at different biological levels, and the addition of chemicals required for efficient production. Each of these requires management strategies and control and regulation. Within Europe, the major areas used are along the Scottish West Coast and island margins, the Irish West Coast and the Norwegian fjords. The areas are in transitional water bodies such as estuaries, fjords, and sea lochs and they require certain facilities such as access, good water quality, and low to moderate water flow (low energy hydrophysical regime). Because of

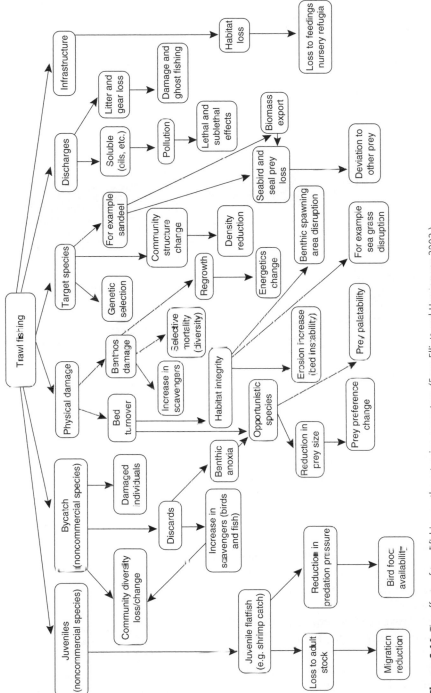

Figure 6.11 The effects of trawl fishing on the estuarine ecosystem. (From Elliott and Hemingway 2002.)

Figure 6.12 A floating sea-cage fish-farm for salmonid fish (salmon or trout). Such farms have become common sights on the fjordic estuaries of Norway, Ireland, Canada, New Zealand, Scotland, and elsewhere. (See also Fig. 6.13.)

the latter, the areas are likely to be stratified. In essence, the effects result from the nature of containing and feeding the fish, for example, salmon are usually fed on pelleted food, composed mainly of fish meal, and up to 20% of this food may not be intercepted and will fall to the bottom, along with the feces produced by the fish (Fig. 6.13). Since the fish cages are usually placed in shallow waters, the waste material rapidly falls to the bottom and accumulates there.

In temperate areas, cages retaining fishes and trays or ropes supporting bivalve shellfish and in tropical areas, cages retaining crustaceans, mainly prawns, are kept within the water column. This produces water column effects through the addition of wastes (as particulate organic matter (POM), dissolved

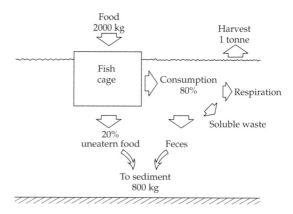

Figure 6.13 The environmental impact of the floating-cage culture of salmonid fish. Values are expressed in terms relating to the production of 1 tonne of fish. (After Gowen *et al.* 1988.)

organic matter (DOM), excess food, feces, mucus, and detritus). The wastes will add nutrients thus having the potential to change oligotrophic (nutrient-limited) systems to more nutrient rich ones. Naturally nutrient-rich systems (hypernutrified) can become eutrophic. There may be a reduction of Dissolved Oxygen (by fin/shell-fish demand and the Biochemical Oxygen Demand of the wastes) and the potential for red-tides (noxious microalgae blooms) has been suggested for the Scottish West Coast. There will be a hydrographic disruption by the structures (they will impede water flow, especially where there is excessive weed attachment) and the inputs from the caged organisms will lead to microbial change through and input of fecal microorganism input. Finally, there may be an increase in suspended solids hence an increased turbidity, leading to potential for a reduction in primary productivity. Many of these effects fall within the generally acknowledged symptoms of eutrophication.

Arguably the greatest impact will be in changes to the sediment and benthos. There will be an accumulation of solid wastes on the bed given that 5–10% of food is wastage and 25–30% of the food weight will be excreted as feces. In the case of bivalve shellfish there will be the biodeposition of sediments through pseudofeces production, hence changing the nature of underlying sediments. The hydrodynamic change because of impeded water flow can lead to sediment structure change through the production of a low energy area. The organic input and enrichment of sediment will produce anaerobic sediments, which may ultimately affect the water quality through the production of hydrogen sulfide (H_2S) and methane (CH_4), which in turn are toxic. Anoxic sediments will affect the water column if the Redox Potential Discontinuity, as the boundary between the upper oxygenated levels and the lower anaerobic levels, migrates out of the sediments. This is then reflected by the development of *Beggiatoa* mats utilizing the hydrogen sulfide. In turn, the anaerobic conditions

lead to the development of opportunist populations and thus benthic prey quality/palatability has been changed for fish and megafaunal predators. Brown *et al.* (1987) have shown that underneath the floating salmon cages the sediment is highly reducing (negative E_h) and azoic. A highly enriched zone, dominated by the opportunistic and pollution tolerant polychaetes *C. capitata* and *S. fuliginosa*, occurred from the edge of the cages out to approximately 8 m. A slightly enriched zone occurred at up to 25 m, beyond which the fauna was unaffected by the cages. Thus the ecological effects of this form of organic enrichment are severe but limited.

In the case of shellfish, farming of mussels has expanded in Spain, New Zealand, and Ireland. The mussels are grown on suspended ropes, feeding on natural planktonic material. In studies of mussel farming in the estuarine Ria de Arosa of northwest Spain, Tenore *et al.* (1982, 1985) have shown that mussels are a "key species" in determining ecosystem structure and dynamics of the whole area. By intensive mariculture, man has replaced the zooplankton with mussels as the dominant herbivores in the area. The major changes are:

1. The surface area of, and detritus from, the mussels support a dense epifaunal community, which utilizes 90% of the mussel feces, and supplies food to demersal fish and crabs.
2. Epifaunal larvae, rather than copepods, dominate the zooplankton community.
3. Nutrient cycling by mussels dampens phytoplankton oscillations and contributes to high seaweed production on ropes.
4. Heavy sedimentation of mussel deposits changes the sediment regime and lowers infauna production.
5. Transport of particulate organics derived from mussel deposits from the farming areas enhances benthic biomass outside this area, and might support near-shore fisheries, especially for lake.

Tenore concluded that the raft culture of mussels affected food-chain patterns and production in generally positive ways, although admitting that the infauna benthos near the farms was typical of polluted conditions. Considering this study along with several others, we can summarize in Fig. 6.14 the possible effects of mussel farming on the estuarine ecosystem. While mussel rafts or ropes reroute the flow of energy or materials, they do not add any extra nutrients to the ecosystem, unlike caged finfish fed on prepared food, and any changes involve processes different from eutrophication caused by an enhanced nutrient supply.

Both finfish and shellfish culture will produce impacts on the indigenous fishes through the alteration of normal habitats, impeding of water flow, and disruption of migration routes. There is the introduction and/or displacement of nonnative stocks

and the introduction of exotic species. In cases where fishes such as salmonids have been reared especially for culture, escapees from the genetic monoculture will produce an alteration in the genetic makeup and a possible reduction in genetic fitness of the natural populations. There may be new and increased disease introduction as well as parasite introduction/concentration. Finally, the removal of locally or distant caught marine fishes for fish meal for feeding the farmed fishes will impact on those stocks.

Given the importance of these areas for birds, aquaculture has the potential for impacts on this use. There may be acute and chronic disturbance of breeding, feeding, and overwintering areas and the disruption of the site (through the loss of land and water usage) as well as an influx of opportunist birds

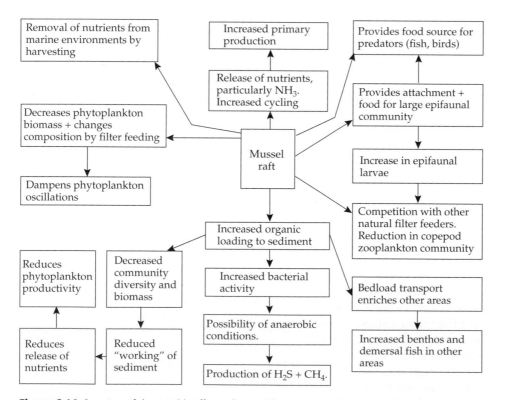

Figure 6.14 Summary of the possible effects of mussel farming. Note that some of the effects are contradictory, and not all effects will be seen at one site. (From Gowen *et al.* 1988.)

feeding. The activities can lead to mortalities through the control by shooting (which requires an exemption to bird protection Acts in certain countries) and there may be accidental or deliberate trapping of birds (through the use of anti-predator nets etc.). These adverse effects may be minimized by site considerations, gear improvement, and the use of scaring devices or sacrificial food, and the adoption of a policy on shooting.

Aquaculture requires the addition of chemicals and these may be given as enteric treatments (by mouth, e.g., with the fishfood) or by immersion treatments. These include food additives such as vitamins, mineral mixes, and pigments, the latter are required in the case of salmonids in order to produce pink flesh, which is acquired in the wild by eating a crustacean diet not used in farming. Anaesthetics and narcotizing agents are required to minimize stress of the fish during handling and vaccines and other therapeutants are administered for health. Disinfectants and antibiotics will be used to minimize microbial effects and, given the increased possibility of external parasite transfer when large numbers of fishes are confined, pesticides (such as Nuvan/Aquaguard (dichlorvos based) and ivermectin) are used to combat infestation of fish lice. Finally, antifouling agents/treatments (such as Cu, TBT, bitumen, and slip-paints) may be used to control macrofouling of the gear.

Aquaculture thus has the potential for impacts at different biological levels. Management strategies may be employed to reduce or remove the effects of aquaculture. The developer may collect the excess solids by funnel or suction system although decomposition, a release of nutrients, and poor water quality can still occur. Pumping or bed trawling to aid breakdown could disperse the wastes. Good siting principles will minimize the effects or the cages may be moved at intervals, with those intervals being judged by the degree of bioturbation under the cages, hence using the assimilative capacity of the area. Feeding techniques using low-wastage feed and hand-feeding, coupled with the type of food (slow sinking, and with a balance controlled to reduce NH_3 input) will reduce the effects. The use of good practice such as the removal and destruction of mortalities, rather than into the water column, and the use of biological control for pests (e.g. lice control using goldsinny wrasse) will minimize biological effects. In addition, fishfarmers and environmental protection agencies determine the potential effects using numerical modeling (based on volume of stock, water flow, depth), coupled with self-monitoring for the health of sediments and benthos. Sampling and/or photography then achieve the assessment. Finally, the use of polyculture, as a mixture of fin- and shell-fish farming, or the use of ranching as compared to cages can be a wiser use of the environment.

6.8 Industrial contamination

Waste from industrial sources enters estuaries from industries located on the banks of the estuary as well as that from inland sites may be discharged into rivers or public sewers, which subsequently arrive in estuaries. The waste from industrial sources may be treated before discharge, so that the effluent arriving in the estuary has a considerably reduced concentration of waste materials, or the waste may be discharged in an untreated form. All industries produce some amount of waste, ranging from large volumes of water, which have been used for cooling purposes, through to chemical waste products which may be extremely toxic even in small quantities.

6.8.1 Petrochemical discharges and inputs

Crude oil is a complex natural mixture dominated by hydrocarbons but also containing non-hydrocarbons and metals, each fraction have a varying toxicity, behavior, and persistence in the environment. It has been estimated that over 2 million tonnes of oil per annum is discharged into estuaries, being approximately one-third of all the oil discharged into the world's oceans. Thus estuaries do indeed receive a disproportionately large burden. Whereas oil spillages and tanker accidents receive a large amount of press coverage, by far the greatest proportion of oil enters the estuaries and coastal seas from diffuse sources such as urban and river runoff or domestic or industrial waste. The effects of the oil industry on estuaries may be divided into (1) the impact of gross spillages, due to shipping accidents or human error at a loading terminal, (2) the impact of effluents produced by refinery and petrochemical industries, and (3) the effects of oil extraction.

Oil spillages

Oil spillages from harbors or collisions have many and varied potential effects, which are the result of the nature of the oil spilled, the characteristics of the estuary and the climatic conditions. Some of the different fractions may be toxic while others are relatively inert and all the fractions will be subject to physical, chemical, and biological degradation. When oil is spilt from a harbor or a collision it reaches water as a separate phase, which is largely immiscible with the water, and generally forms a surface slick (Fig. 6.15). The slick will spread and be subject to physical and chemical breakdown through photo-oxidation, evaporation and, during high winds, aerosol formation. Hence the importance of the weather conditions affecting the fate and effects of the spilled oil. Many of the lighter and most toxic elements can be lost by weathering, but unfortunately, in the confined space of an estuary there may not be time for this to occur before the spilt oil is deposited on the shore.

The surface slick and the sinking oil will be transported to areas of the estuary depending on the wind and tidal conditions and the nature of the estuary. For example, a spillage at the mouth of the estuary will impact on local rocky shores and sandy beaches whereas that pushed up the estuary could affect mudflats, salt marshes, and reedbeds. Once on the sandy beaches, the oil can be pushed into the coarser sediment by tidal pumping and once in the sediment, especially if there is no oxygen penetration then aerobic degradation will be reduced. Hence the oil can remain for many months. On the shore the oil acts mechanically, smothering animals in burrows or on rocks, and excluding the light from plants. Given that mudflats have larger infauna and predator populations than sandy beaches then oil deposited on the former has a greater smothering effect and affects predator–prey relationships. Oil deposited on higher energy sandy beach surfaces and on rocky shores will be subjected to greater physical degradation and thus may breakdown more easily. In contrast, oil deposited in low energy estuarine environments such as mudflats, seagrass beds, salt marshes, and reedbeds will persist for longer and is also more difficult to remove by mechanical methods. It will infiltrate creeks and between the plants and thus be protected from mechanical breakdown. On a salt marsh the soiled parts of the plant die, but if it can survive this, the plant can regrow.

The oil will also be mixed with the water column forming first water-in-oil emulsions and then, with increasing mixing, oil-in-water emulsions (often called chocolate mousse due to the color and consistency). Zooplankton will ingest the oil droplets and produce fecal pellets, which can aid breakdown by bacterial biodegradation on the large surfaces. These droplets and pellets may then settle out of the water column. The water-soluble or more volatile compounds are most toxic, and spillages in American estuaries of fuel-oil containing 45% low aromatic oils have devastated commercial shell fisheries. The heavier compounds within the oil, which remain after evaporation, will tend to sink, where they undergo microbial decomposition by many microorganisms, which are capable of degrading petroleum hydrocarbons as well as naturally occurring biogenic hydrocarbons.

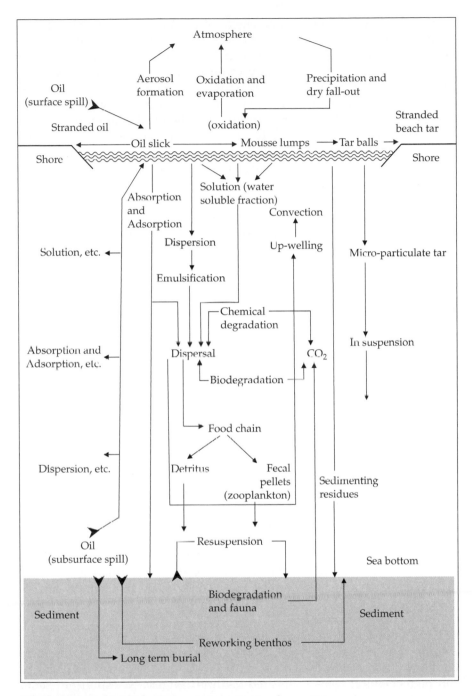

Figure 6.15 The fate of oil spilled or discharged into surface waters (modified from CONCAWE 1981, www.concawe.org).

The most resistant oil fractions, the heavier and less volatile components, however, remain as tar-balls for many weeks if not months.

Given the serious consequences of oil penetrating the more delicate habitats in estuaries, especially those where oil is difficult to remove once it settles, then the most common advice on how to deal with oil spills in estuaries is to contain the oil by a boom if possible, and then to use mechanical removal with pumps for the trapped oil. If it comes ashore on sandy beaches then mechanical removal with shovels or excavators is the preferred mode of treatment, rather than emulsifiers if at all possible. The earlier emulsifiers, or detergents, used in clean-up campaigns to disperse oil were often more toxic to estuarine life than the original oil. Newer emulsifiers are less toxic. Oil washed off rocky shores with hot and/or freshwater can cause greater community effects than the oil if left alone. Oil settling on to mudflats, salt marshes, sea grass beds and reedbeds in estuaries cannot be removed without considerable damage to those habitats and so if these areas are oiled then the decision is very often to leave the oil to natural clean-up. Consequently, an estuarine oil-spill contingency plan may involve recovering the oil from sandy beaches rather than other habitats, especially as the beaches may have good access and will support heavy machinery. Figure 6.16 shows the recommended decision processes involved in dealing with an oil spill within an estuary.

Petrochemical discharges

Less dramatic, but potentially more dangerous to estuaries than oil-spills, are the insidious effects of the continual discharge of industrial effluents from petrochemical complexes. The effluent from such industries is mostly hot freshwater, which may contain certain amounts of chemicals, including hydrocarbons (oil) from refineries, or chemical waste products such as phenols or ammonia from chemical works. A refinery effluent may contain only 10–20 ppm of oil, which is difficult to remove and seen perhaps only as cloudy water with an oily surface sheen, but as a flow of 7.58 cumecs this produces more 6.55 tonnes of oil per day. The impact of effluent from petrochemical complexes in the Medway and Forth estuaries was examined before major clean-up operations. In the vicinity of the outfall an abiotic zone was found, with no life at all. Beyond this was a "grossly polluted zone" in which only Oligochaetes were found, in small numbers. Then followed a "polluted zone" with abundant Oligochaetes and *Manayunkia* (a small polychaete), and occasional specimens of other species such as *Hydrobia*. Finally came the "largely unpolluted zone" at over 1-km from the effluent, with *Macoma, Cerastoderma, Nereis, Nephtys*, and *Hydrobia* often abundant, and fewer Oligochaetes. It may be seen that the zonation of species and their abundance is remarkably similar to that noted above for the effects of organic enrichment. Since oil is a biological product derived geochemically from organic material, it is perhaps not so surprising that the effects of oil on estuarine ecosystems are rather similar to the effects of other excess organic materials. Throughout the world oil refineries have made strenuous efforts to reduce their waste effluent, and modern oil refineries have considerably less impact on the estuarine ecosystem than their older counterparts. When an oil refinery discharge is reduced, or terminated, then the resilient estuarine fauna can generally recolonize the area rapidly.

Oil and gas extraction

Oil is extracted from several estuarine locations, most notably the Nigerian oilfields are located within Nigeria's estuarine delta region, and blow-outs from the wells have occasionally led to environmental problems. The shallow areas of eastern England and the Dutch coast border on the southern North Sea gas fields and the large estuarine area

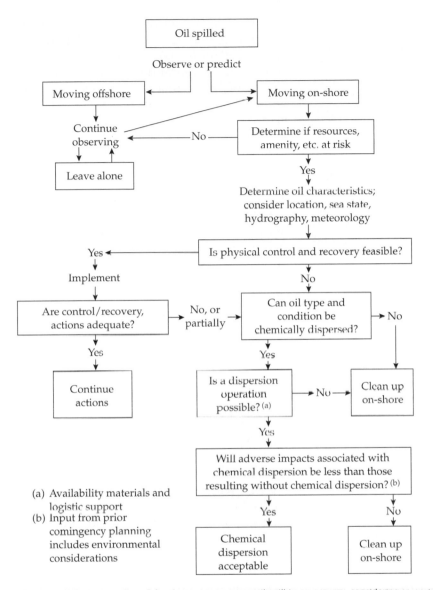

Figure 6.16 Decision "tree" for the response to an oil spill in an estuary, considering in particular whether a chemical dispersant is acceptable. (From Concawe, 1981.)

Morecambe Bay in NW England also sits on a gas field and thus these areas support gas rigs as well as seabed collecting systems. Gas and oil come ashore by pipeline into several estuaries, and so formation water and production water has to be treated and disposed. Formation water is the fossilized water trapped with the oil and gas, which may contain high levels of minerals and be at a high salinity whereas production water is that pushed down into ageing wells to increase production. Each of these will contain oily residues and thus require treatment before disposal into the estuary.

6.8.2 Heavy metals

The effects of other chemical waste products, such as heavy metals, may be quite different to the effects of hydrocarbons. Particular attention has been paid to discharges of heavy metals such as mercury, cadmium, copper, iron, lead, zinc, or chromium, from a variety of industries including chemical plants, metal-processing industry, and mine waste water.

Two approaches have been used. First, examination of the toxicity of the metals to estuarine organisms, and second, examination of the metal concentrations within organisms. A comprehensive review of the effects of temperature and salinity on the toxicity of heavy metals to

estuarine invertebrates by McLusky *et al.* (1986) has shown that toxicity values determined under fixed (or single) temperature and salinity regimes are inappropriate for evaluating the effects of environmental factors in modifying the toxicity of metals to estuarine animals (Figs 6.17 and 6.18). Indeed, the range of toxicity values for a single estuarine species exposed to different temperatures and salinities may exceed the entire range of toxicity values previously published for a wide variety of marine or freshwater animals. A rank-ordering of the toxicity of metals is: mercury (most toxic) > cadmium > copper > zinc > chromium > nickel > lead and arsenic (least toxic). A taxonomic order may also be seen of Annelida (most sensitive) > Crustacea > Mollusca (least sensitive). For all estuarine animals, heavy metal toxicity increases as salinity decreases and as temperature increases. From a variety of studies it appears that heavy metals may compete with calcium and magnesium at cellular uptake sites and thus disrupt the key physiological process of osmoregulation, which is so vital for survival in estuaries.

Concentrations of metals below those needed to cause acute toxicity may cause a variety of sublethal effects, such as inhibiting reproduction or growth, and will also be accumulated within the tissues of the organisms exposed to the metals. The metal content will generally increase with size and age as the metals are accumulated, and are unable to be excreted. Simple measurements of the total concentration of metals within estuarine organisms may give misleading impressions of the impact of the metals on the organism. Within estuarine molluscs, for example, the formation of vesicles within cells, which enclose the metal within a membrane, has been found. These vesicles prevent contact of excess metal with vital constituents and effectively detoxify the metal. Estuarine annelids, such as *Nereis* have been shown to be able to detoxify metals by accumulating them within their jaws. Although detoxification may prevent a mollusc or an annelid from being

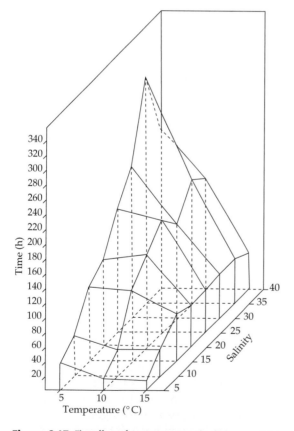

Figure 6.17 The effect of temperature and salinity on median survival time (h) of *Corophium volutator* at a chromium concentration of 16 mg l^{-1}. (After Bryant *et al.* 1984.)

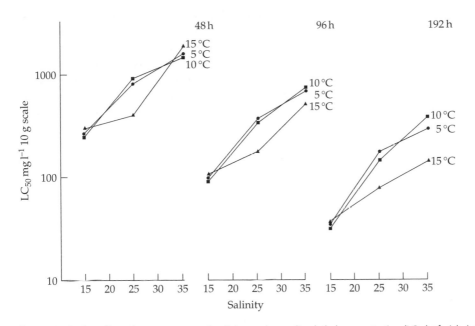

Figure 6.18 The effect of temperature and salinity on the median lethal concentration (LC_{50}) of nickel to *Macoma balthica*. (From McLusky *et al*. 1986.)

adversely affected by excess metals, a predator eating it may not possess the same ability and may be affected by the total concentration of the metal. Unlike some other pollutants, heavy metals may show *biomagnification*. That is, they show an increase in concentration as one proceeds up the trophic levels of the estuarine ecosystem. For that reason, humans should always avoid eating any shellfish exposed to heavy metals.

The processes of sedimentation within estuaries, discussed in the first chapter, may serve to relieve the estuarine and marine ecosystem of some of the worst effects of heavy metal pollution. As metals are discharged into the turbid waters of estuaries they may become rapidly bound onto the surface of the fine sedimentary particles. As the sedimentary particles settle on to the intertidal mudflats, the metals are gradually buried. In many of the estuaries bordering the North Sea, it has been estimated that about half of the metals entering the estuary become trapped within the estuarine sediments, and only a lesser amount is eventually discharged to the sea. If estuaries are to serve as such vital traps for pollutants, it is essential that the natural intertidal mudflat areas remain intact and undisturbed. In several cases it has been shown that man's activities may disturb such metal burial, for example, dredging will often liberate heavy metals from sediments, and even bait-digging for worms has been shown to lead to an increase in the total concentration and bioavailability of metals.

6.8.3 Organohalogens

The addition of pesticides and organohalogen compounds to estuaries poses considerable threats to the estuarine organisms. The organohalogen compounds are the polychlorinated biphenyls (PCBs), including hexacholorobenzene (HCB), lindane, hexachlorocyclohexane (HCH), dieldrin, and dichloro-diphenyltrichloroethane (DDT) used both as pesticides, and for industrial purposes (especially in electrical equipment). Unlike petroleum hydrocarbons chemical oxidation or microorganisms cannot degrade these materials, generally known as the halogenated hydrocarbons (hydrocarbons containing chlorine, bromine, iodine, or fluorine). Thus, like metals, these substances are permanent additions to the estuarine ecosystem, but, compared to metals, even minute quantities of these man-made substances can pose a hazard to estuarine life. Any animals, which ingest them, cannot excrete these substances, and they thus accumulate within their bodies, principally in fatty tissue although detoxification can occur. Food-chain biomagnification can occur so that top predators have the highest concentrations, and may suffer most. Until the effect of these substances was realized, and their production curtailed, about 1 million tonnes of PCBs were produced around the world. Much of this material has ended up in estuarine and marine organisms, and will continue to circulate through estuarine ecosystems.

Pesticide or industrial PCBs in very low concentrations, for example, $1 \ \mu g \ l^{-1}$ (=1 ppb) have been shown to be lethal to estuarine life both in laboratory tests, and in field mortalities around industrial and agricultural discharges containing pesticides. Due to the biomagnification effects of PCBs, top predators such as birds and seals are particularly at risk. Within the Baltic Sea, for example, high levels of PCBs within seals have been held responsible for the failure of reproduction in many female seals.

6.8.4 Radioactive discharges

The subject of the addition of radioactive substances to estuarine environments arouses considerable public concern. Man has, during the latter part of the twentieth century, added radioactive material to seawaters from the fallout of atmospheric bomb testing, from waste discharged from nuclear power stations and reprocessing plants, and from the use of radioactive material in submarines. The unique detectability of radioactive isotopes enables the anthropogenic additions to the environment to be identified clearly and quantified. It is, however, often forgotten that the man-made additions may be very small in comparison to the amount of radioactivity that naturally occurs in seawater. Over 90% of the natural radioactivity of seawater arises from potassium-40, almost 1% from rubidium-87, and the remainder is virtually all from the isotopes of uranium. The concentrations of these elements vary proportionately with salinity, so the natural radioactivity of estuarine water is governed by salinity, with greater concentrations in saline waters and lesser concentrations in brackish waters. At a typical salinity of 33 the natural radioactivity has been measured as $12 \ Bq \ L^{-1}$ (1 Bq (Bequerel) is one disintegration per second). In the Forth estuary and the Firth, the volume of water is 20 km^3 or 2×10^{13} l, so Leatherland (1987) calculated the total natural radioactivity of these waters as 240,000 GBq, indicating a total of 60 tonnes of dissolved uranium in these waters. In this area a naval dockyard adds 7 GBq per annum, and low levels of isotopes will be added in sewage systems derived from hospital or laboratory sources. A nuclear power station has added a further 200 GBq per annum, but it can be seen that these sources are minute compared to the natural levels of radioactivity present.

In Britain and France the major anthropogenic source of radioactivity is from nuclear reprocessing plants at Sellafield and La Hague. These plants recover uranium for re-use from the fuel rods of nuclear power stations, and discharge large quantities of waste water with a low radioactive content, but because of the volumes involved they generate a considerably greater input of radioactive material than nuclear power stations or other man-made sources. Some of the waste products decay rapidly or are adsorbed onto fine sediments. The fine sediments can be carried into local estuaries, and in the case of Sellafield these need to be regularly monitored. The isotope caesium-137 remains in solution, and since it does not occur naturally it serves as an effective "marker" for the Sellafield waste. Regular annual surveys, of British coastal waters, show that waste from Sellafield is predominantly carried north and west away from the discharge, and after passing around the north of Scotland enters the North Sea and its estuaries. Close to the discharge point the concentration of caesium-137 is 0.5 Bq kg^{-1}, and declines to 0.05 Bq kg^{-1} in the North Sea (Fig. 6.19). These concentrations are considerably less than the natural radioactivity levels of seawater, but are greater than any other anthropogenic sources of radioactivity. In recent years the quantities of waste discharged from reprocessing plants has diminished considerably, but they remain the main source of man-made radioactivity in estuaries. Laboratory studies have shown acute lethal effects of high radiation doses on many organisms, but it must be said that existing levels of natural and man-made radiation in estuaries and the sea have so far produced no measurable environmental impact on these ecosystems.

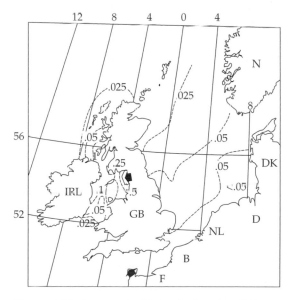

Figure 6.19 Concentration (Bq kg^{-1}) of caesium-137 in filtered water from north-east Atlantic waters, 1987. The arrows show the discharge points of material from nuclear reprocessing plants. (Based on data in Hunt 1988.)

6.8.5 Threats from industrial pollutants

Of the various pollutants discussed, three groups of substances pose particular threats to the estuarine ecosystem. These are heavy metals, PCBs, and radioactivity. All these substances are persistent, and cannot be degraded, although they show some slow reversibility due to burial in sediments. All show bioaccumulation, and in the case of PCBs also show clear biomagnification. All show effects on the fauna at low concentrations, often at concentrations well below those shown to be toxic in standard 96-h toxicity tests. The threats posed by metals and PCBs can be demonstrated in present estuarine ecosystems, but radioactivity effects have not been shown to disrupt ecosystems so far. By contrast, organic materials and oil may show severe effects in the vicinity of discharges, but these materials can be biologically degraded, and their effects can be completely reversible at the estuarine ecosystem level. So, when considering pollutants in estuaries one must consider not only the quantity and quality of the waste, but also whether it can be degraded and whether its effects are reversible.

6.9 Land-claim, coastal defences, and engineering works

Land-claim typically represents only gains to agricultural or industrial land users, and losses to the wildlife and fisheries of estuaries. Against these losses may be set the occasional gains of lagoons or other wetlands. For many estuaries the destruction of habitat due to various land-claim schemes has had a far greater impact than any of the effects of any polluting discharges entering the estuary. Estuaries are areas of natural deposition of sediments, and as salt marshes develop they slowly create new areas of dry land. Neighboring farmers seeking to extend their acreage often acquire these new areas of dry land. Throughout the world, for many centuries, man has drained salt marshes and with the protection of sea walls converted them into productive agricultural land. In Britain such land-claim has been predominantly on the east coast, and in Denmark such land-claim has occurred in West Jutland. In India and Bangladesh such claimed land has proved to be particularly suitable for rice cultivation. Such processes of agricultural land-claim are, however, gradual in comparison to the large-scale land-claim schemes undertaken to meet the needs of industry. Modern industrial developments seek areas of flat land close to available transportation, and as other land may not be readily available, large-scale estuarine land-claim is often suggested. For such a land development scheme, the first stage is usually the construction of a seawall or bund across the intertidal mudflats, followed by the infilling of the bounded area with dredged estuarine mud, or hard-fill material derived from quarries, mine waste, ash waste from coal-fired power stations, or domestic refuse. In other schemes, the impounded area may be filled with freshwater to form reservoirs, and especially in the Netherlands the impounded area may be drained and converted into agricultural land or polders.

Whatever the use of the claimed land, it represents a total loss to the estuarine ecosystem. In the estuaries of eastern Britain, for example, industrial land-claim has obliterated over 90% of the Tees estuary, leaving less than 10% of the original intertidal area (Fig. 6.1), and in the Forth estuary, approximately 50% of the intertidal area has been lost through land-claim in the past 200 years (Fig. 6.20). At first such land-claims were for agricultural purposes, followed by harbor works, and more recently for industrial purposes, including sites for power stations, refuse disposal and fly-ash dumping. Within Southampton Water land-claim has taken place over many years to meet the needs of new docks, oil refineries, power stations, and other industries (Fig. 6.21). While the salinity regime has been hardly affected by the land-claim schemes, the removal of extensive salt marshes has reduced substantially the input of organic detritus to the ecosystem, and made the whole ecosystem more dependent on phytoplankton production. As intertidal areas have been obliterated, the commercial clam harvest has fallen from 100 to 60 tonnes per annum, and the bird population feeding intertidally has declined substantially.

San Francisco Bay is the largest estuary on the Pacific coast of the United States, with a water surface area of 1240 km^2, and probably the most modified estuary in America. Large areas of wetland, or salt marsh, have been destroyed by land-claim, so that only 6% of the original 2200 km^2 of wetland remain.

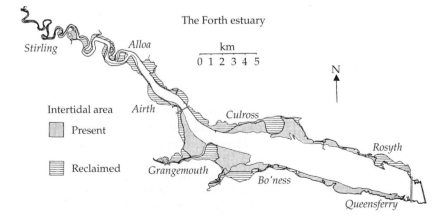

Figure 6.20 Land-claim of intertidal areas of the Forth estuary in Scotland for agricultural and industrial purposes over the last 200 years. (From McLusky 1987.)

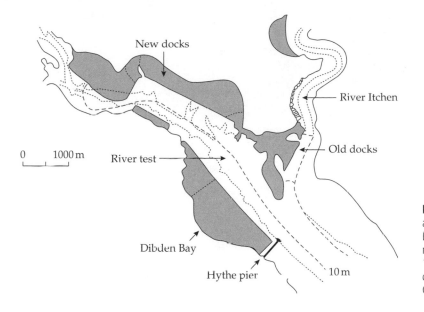

New docks

River Itchen

0   1000 m

River test

Old docks

Dibden Bay

Hythe pier

10 m

Figure 6.21 Claimed for land areas around Southampton Water, southern England. The fine dotted line represents the low water mark of 1830, and the firm dotted line the current dredged channel. (After Coughlan 1979.)

Due to inland water engineering schemes less than 40% of the original river flow now reaches the estuary, a factor that has considerably reduced the flushing of pollutants from the Bay area. On the east coast of the United States more than 25% of the estuarine wetlands have already been lost through land-claim. Wherever, and for whatever purposes, land-claim occurs, the birds and fish, which feed on estuarine mudflats, simply find that a large proportion of their food supply has disappeared. American legislation now requires that new land-claim schemes must compensate for the loss of habitat by the provision of alternative localities; however, it is very doubtful if such compensation can ever fully replace a habitat which has taken centuries to develop.

6.9.1 Dutch delta works

Faced with the problem of the threat of large-scale flooding from the North Sea, the Netherlands has claimed extensive estuarine areas. In the past three decades major changes have occurred in the estuaries of the southern Netherlands, in the "Delta Plan," whereby the estuaries of Rhine, Meuse, and Scheldt have been dammed and converted from tidal estuaries into freshwater lakes, brackish lakes, or sea areas with restricted tides (Fig. 6.22). The southernmost estuary, the Westerscheldt, remains open, but the Oosterscheldt estuary has been cut off from its freshwater supply (the Rhine/Meuse), large areas of shallow wetland have been claimed for land, and an 8-km wide tidal storm surge barrier restricts the inflow of seawater. As a consequence it is now a sheltered marine bay, not an estuary. The Grevelingen estuary has been closed at both its river and seaward ends to form a semi-stagnant nontidal brackish-water lake, with a stable salinity of about 1–8. The small Veere estuary has also become a brackish-water lake with a salinity of 7 in the winter, rising to 12 in the summer. The Haringvliet estuary has been closed at its mouth, preventing the entry of salt water, but still permitting the exit of freshwater. Consequently it has become a freshwater lake, fed by the Rhine. The necessity for the Delta Plan to the safety of the Netherlands is clear, but the consequence has often been the obliteration of mussel beds and the elimination of feeding sites for birds and fish.

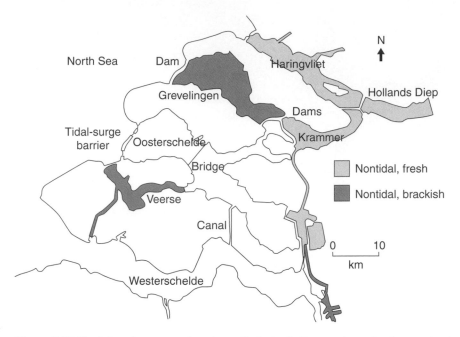

Figure 6.22 The Delta scheme area of southwest Netherlands. The map shows that four previously open estuaries have been partially or completely closed off from the sea. In addition inland compartment dams have created freshwater lakes from previous estuaries. Only the Westerschelde remains as an estuary. (After Nienhuis and Smaal 1994.)

6.9.2 Sea-level rise and estuarine development

In many areas, such as the North American east coast and the European northwest coast, the level of the sea is rising and the intertidal areas are in retreat. Where there is no restriction to that movement at the upper tidal areas then the intertidal can migrate landward although in most of these areas, which are developed, then a seawall will restrict this movement. This is termed "coastal squeeze."

In some northern areas, the land is sinking due to "isostatic rebound," the result of the weight of ice in the last Ice Age being removed from parts of the Earth's crust to the north of these areas. For example, while Scotland and Scandinavia are rising following their land-mass being depressed in the last Ice Age, the land is sinking in the southeast of England and parts of Germany and the Netherlands. Coupled with sea-level rise through global climate change results, for example, in there being a relative sea-level rise of 6 mm in SE England, the site of many estuaries. This change is leading to the erosion of shorelines, especially the almost flat intertidal areas and wetlands of estuaries, and it has been estimated that as a result Louisiana is losing about 40 ha of wetlands every day, or about 15,500 ha year^{-1}. On the Thames estuary in London, England, a tidal barrier has been built to protect the city against very high spring tides which, with the subsidence in land level and rise in sea levels, threatened to flood this major city. Elsewhere sea walls will have to be built or increasingly raised to protect coastal communities. For the estuarine ecosystem, these changes are yet another environmental factor, which will change many of the estuaries and wetlands that we see around us today.

Faced with these threats, society has the choice to build bigger and costlier seawalls, which will exacerbate the effects of coastal squeeze, or to allow the estuary to reoccupy some of the areas that previous generations have removed by land-claim. This reoccupation, known as "managed retreat" is being trialled at several locations and may represent the future for many estuarine areas.

6.9.3 Ports, navigation, and dredging

The use of estuaries for transportation requires harbor works, which can range in size from a single pier to a vast container or bulk-cargo terminal. A single pier may cause localized change in current pattern and sedimentation, whereas a large terminal will have a similar impact to other major land-claim schemes. The major ports of Southampton, Rotterdam, Antwerp, and Hamburg are all situated on major estuaries and all have developed by building on land-claim, building wharves into the deep channel. Estuaries have port developments, which depend on the position of the channel, for example, the four ports of Hull, Grimsby, Immingham, and Goole on the Humber estuary, which are the largest UK ports complex, each developed where the deep water channel touched either the north or the south banks in an otherwise shallow estuary.

In order to maintain access for shipping, many estuaries are dredged. At the dredge site, the substratum is removed along with any fauna it contains, and the dredge spoil is then transported to a disposal ground, usually at the mouth of the estuary, or outside it. The fauna of the dredged and disposal grounds may be affected by being moved or smothered, and pollutants accumulated in estuarine sediments, which act as a sink, may be released. Harbor sites may also be a cause of pollution, especially from oil, through accidents as ships are loaded and unloaded, or due to the risk of collision, or from leaks from ships' engines or bilges. Ship repair and demolition sites also present potential hazards for considerable further "accidental" pollution.

Dredging operations in estuaries can be divided into either maintenance or capital dredging. Maintenance dredging is the periodic and often repeated removal of material, typically sand, silt, and gravel deposited in low energy, accreting areas by river discharge, tidal currents, or wave action in areas previously dredged. In contrast, capital dredging is the first time removal of sediments, which may be necessary to create a new channel, build a new bridge or lay a new pipeline of telecommunications cable. Hence it is likely that sediments for maintenance dredging are recently deposited and relatively un-compacted although they may be accumulating organic matter and pollutants, which easily settle, flocculate, or precipitate from the water column.

The nature of the material, the area, and the type of equipment used dictate the environmental impacts of dredging and dredged material disposal. The different types of dredger include those that remove material mechanically by means of a grab or bucket, whereas suction dredgers transport the material as a suspension via a pipeline. Silts and fine sands can also be cleared from harbors and dock entrances by agitation dredging, that is, resuspending the material and then relying on water currents to carry it away. Usually the dredged material is loaded into a hopper, either in the dredging vessel or in a separate vessel, for transport to the disposal sites where the material is discharged by pipe or bottom-opening doors. The environmental effects thus result from the way the sediment is obtained at the dredge site and liberated at the disposal site.

Disposal sites were first selected for convenience in being close to the dredged area but sufficiently far away so as not to affect navigation. The disposal site for a small fishing harbor or marina may receive maintenance dredging on an annual or bi-annual basis, consisting of a few thousand tonnes of marine sand moved into the harbor area from the adjoining seabed by winter storms. In contrast, the disposal ground for large estuarine ports may receive many millions of tonnes of low density silt several times each day.

In the United Kingdom in 2000, 10 sea disposal sites received >1 million wet tonnes of dredged material, 24 sites received 1 million–100,000 wet tonnes, 40 sites received 100,000–10,000 wet tonnes, and 21 sites received <10,000 wet tonnes. While a relatively small number of sites accounted for a large proportion of the United Kingdom's 2000 dredged material disposal, at least 95 sites had some material deposited in them. A number of the sites received material from several different dredging areas. Despite this, not all dredged material is disposed of at sea as significant tonnages are placed ashore in lagoons, land-claim areas, and landfill sites. On-shore disposal may be used for infilling development land and as beach nourishment for coast protection. In cases where the material being dredged is deemed to be too contaminated for sea disposal then it may be placed in containment sites such as the Slufter system created by the Port of Rotterdam in the 1990s.

Both dredging and disposal of estuarine sediment can affect the physical, chemical, and biological features of the water column and the seabed and the impacts could be short-term or long-term. The potential impacts could include a deterioration in the overall health/quality of the marine ecosystem; a reduction in the socioeconomic aspects of the sea, including fishery and amenity interests; an interference with the legitimate uses (i.e. those legally permitted) of the sea/recreation and navigational aspects, and a reduction in the aesthetic qualities associated with the area.

The potential effects of dredging and dredged-material disposal can be regarded as a set of bottom-up causes, in which the physical system (both in the water column and on the bed) is altered and in turn affect the health of the biological system. The eventual effects on the biological system and its uses by Man can be regarded as a set of top-down responses, for example, the effects on the higher levels of the ecological system (such as fishes, seabirds, and marine mammals) as well as on fisheries and conservation objectives (Elliott *et al.* 1998). As with many of the activities considered here, we can use our knowledge of these effects and the linkages between the different responses to create a conceptual model which, by the nature of the system and the potential changes to dredging and marine disposal, is naturally complex (Figs 6.23 and 6.24, adapted from Elliott and Hemingway 2002).

Dredging will alter the bed topography and bathymetry, which in turn will remove the bed organisms and it substratum, as well as changing the overall hydrodynamic regime of the area. The structure and functioning of those bed sediments and the overlying hydrographic regime (water currents, tidal circulation, etc.) will be intimately linked to the structure and functioning of the bed biological community, principally the invertebrates (Elliott *et al.* 1998). In turn this will influence the fishes and, in estuarine nearshore areas, the birds feeding on those invertebrates. The second major effect of dredging will be the resuspension of the bottom sediments and its effects on the water turbidity, and the liberation of any materials contained and sequestered within the sediments. Dredging will result in the sediments being exposed to aerated waters and this change in chemical nature can re-release the contaminants. In addition, anoxic and organic rich estuarine sediments may generate methane and hydrogen sulfide, which is liberated on disturbance by dredging. The release of those materials into the water column will then have the potential for a biological effect. All of these effects have the potential to influence the fisheries and nature conservation value of the area (Fig. 6.23).

The disposal of dredged material will similarly have the potential to affect the water column, the bed conditions and their biota (Fig. 6.24). Reductions in water clarity through an increased turbidity will in turn affect the primary production by the phytoplankton. The release of any materials contained within the dredged material, either as the water

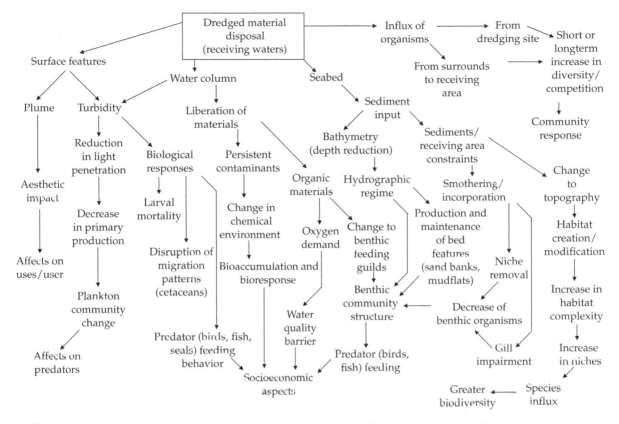

Figure 6.23 Potential environmental impacts of estuarine dredged material disposal—conceptual model. (Modified from Elliott and Hemingway 2002.)

soluble fraction or the release of particulate materials may be the result of the changed chemical environment, that is, anoxic fine sediments liberated into the oxygenated water column will cause the release of pollutants previously sequestered due to the anoxic chemical conditions. Similarly, any organic matter in the sediment will create a water column oxygen demand. The deposited sediment will change the nature of the bed sediment, if it is of a different particle size and it can have a smothering effect on the bed community as well as bringing new organisms to an area. Both of these features will affect the structure of the bed community and in turn the demersal and benthic fishes feeding on that bed community.

Whereas most of the above impacts of dredging and dredged-material disposal relate to the ecological system, the resultant impacts on the uses and users of the marine environment are often of greater public concern. These include the actual or perceived effects on socioeconomic aspects such as fisheries, and aesthetic aspects, including recreation and tourism. Similarly, the perceived or actual effects on the conservation importance of an area will be of concern, especially where the habitats and species within and adjacent to the dredging and disposal areas are of importance.

In addition to capital and maintenance dredging, estuaries are often the sites of the

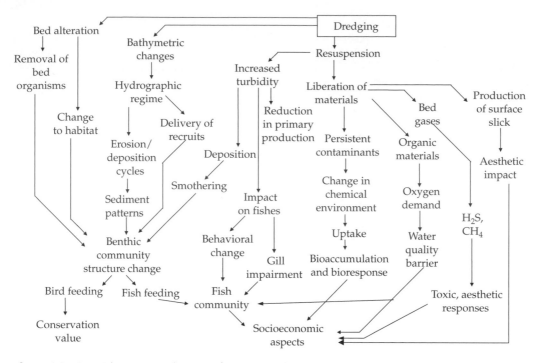

Figure 6.24 Potential environmental impacts of estuarine dredging—conceptual model. (Modified from Elliott and Hemingway 2002.)

winning of aggregates to supply the needs of construction industries. Sand and gravel are often dredged from the bed of the lower reaches of an estuary. However, given that bed sediments are rarely a single type of material, the desired grade of sediment requires to be separated from other, often finer material and the latter is then discharged over the side of the dredging vessel. This can lead to smothering of the bed sediments and fauna and changes to the water quality, especially the turbidity levels. However, given that estuaries are often highly turbid and their biota are adapted to sedimenting conditions then the effects may be less than on the open coast.

6.10 Power generation

Estuaries have long been the sites of power generation, from having small-scale water wheels operated by tidal flow to grind grain, to

now supporting large power plants and having barrages placed across the estuary for large-scale tidal power. Power generation can be separated into conventional plants and alternative sources. The former use fossil fuels, such as oil, gas or coal, or nuclear energy to heat water to create steam, that turns turbines and need large volumes of water to condense the steam after use. In contrast, alternative sources, often called renewable generation, uses the tides, waves, water currents, winds, the sun, or freshwater impoundment (hydroelectric).

Many of these generating sources occur on estuaries or at least impinge on estuarine processes, especially conventional plants and tidal power systems. Wave power is not successful in sheltered estuarine areas and as yet there are no water current systems in which an open propeller is placed within the water body. Hydroelectric schemes are not found in estuaries, despite this these schemes may affect the inhabitants of estuaries such as impeding the migration of diadromous fishes, such as the shad (*Alosa* spp.) or the salmon (*Salmo salar*). Conventional power generating plants are often sited in estuaries because of their associated infrastructure and the availability of a large supply of cooling water. As with all major structures, the power plants will have an impact on the estuarine environment but also the estuarine environment has an effect on the power plant.

The initial and often long-term effects are from the construction and site development and maintenance. Land-claim/land loss, especially of intertidal area, may occur to provide space for construction or for the waste from the power plant. Any change to the shape of the estuary will produce hydrophysical disruption, with changes to the hydrography, sediment regime, and bathymetry. As with any major construction, there can be construction discharges and runoff of oils, solids, and litter. In the case of large power plants there will be dredging, seabed aggregate extraction, aggregate movement, and storage as the seaborne delivery of building materials is often the most efficient. The creation of nuisance, often termed aesthetic pollution, from light, noise, and vehicle and ship movement, will affect the large overwintering wading bird populations. In addition there will be hindrance to resource exploitation by boat and fixed net fishing and to recreation and tourism. Finally, the visual impact of the plant may affect the amenity value of the area.

The operation of power stations and production of power causes environmental impacts both from the waste, especially from coal-fired power plants, and because of the intake and discharge of cooling water. Coal-fired power plants use pulverized coal as fuel and thus they produce pulverized fuel ash or fly ash (PFA), which may be up to a million tonnes per year for a large power plant. In some cases this material has been used to reclaim estuarine intertidal areas, such as at Culross Bay, Scotland, thus removing productive wader feeding areas, whereas in other cases the material has in the past been dumped at sea. Such a fine material input changed the nature of the bed sediments by blocking pore spaces, smothering the bed fauna, and affecting colonization. Trace metals in the ash can also leach into the surrounding waters. This material has been used in producing "breeze blocks" for construction and sometimes for creating artificial reefs, such as the experimental site near Southampton, southern England.

The greatest environmental impacts are, however, due to the main reason for siting the power plant on the estuary the intake and hence discharge of large volumes of water for cooling the condensers. This is referred to as direct cooling and thus removes the need for cooling towers common to inland power plants. These impacts can be separated into those related to the intake and those for the discharge. A large power plant may require 30–50 $m^3 s^{-1}$ (cumecs), the equivalent of a medium sized river, and so the power plant has to be positioned where there is an adequate water source, hence an estuary. In taking in this volume, large and small material is sucked into the plant, termed impingement. Material greater than 1 cm^2 will be retained on the initial rotating screens inside the power station. This material includes fishes, large and mobile invertebrates such as lobsters, crabs and, in the outer parts of estuaries, squid and octopus, seaweed, jellyfishes, and litter. The amount and nature of this material depends on the type of biological communities adjacent to the plant and the position of the intake pipes.

Up to several tonnes per day of fish, other organisms and debris can be impinged, especially if large inshore and estuarine fish migrations occur. For example, a power station in the Forth estuary, Scotland, is situated in an area supporting large overwintering estuarine fish populations (Elliott *et al.* 1990), which become impinged. Hence power plants have been described as stationary trawlers. The impinged fish suffer injury and death through the pressure effects, which cause swim-bladder damage in round fish, external damage to scales and fins, for example, in delicate fish such as clupeoids, and the hemorrhaging of eyes. If there is a high mortality then this might have an influence on the inshore and estuarine nursery function and on migration routes, hence disrupting fisheries. Such an influence is greater on round-fish than flatfish. Large amounts of impingement, such as of jellyfish in the Emshaven power plant on the Ems Estuary, the Netherlands, can also severely affect the operation of the plant and even lead to the plant being shut down. The impinged material, termed trash, can be returned to the estuary or disposed to sea or landfill under licence although power plants would generally prefer to reduce the impinged levels by the use of intake design or acoustic, light, or bubble screens.

Materials smaller than $1 \, cm^2$ will pass through the rotating screens and thus pass into the main cooling system of the plant. This small material includes larval forms as well as the silt found in high concentrations in the estuarine waters. This leads to the fouling of the pipes inside the plant by sessile fauna (spat, barnacles, saddle oysters, bryozoans, hydroids, and tubeworms) and bacterial film. If left unchecked this leads to blockages and a reduction in the cooling efficiency, especially condensers and areas within the cooling water stream, any auxiliary stream and, in the case of nuclear stations, the reactor cooling water. As such there is the need for biofouling prevention measures involving biocide, antifouling, and anticorrosion treatment. The latter relies on the transport, production, storage and use of biocides from oxidizing compounds such as liquid chlorine, powdered sodium hypochlorite, electrochlorination, and bromine-enhancers. There will be large mortalities of planktonic organisms and larval stages, although there are vast quantities of these in the water column. The introduction of halogenated antifouling agents coupled with high levels of organic matter in the water and the

heat produced in the system leads to the production and liberation of chlorine residuals and organohalogens (bromoform, bromoamine, chloroform, chloroamine, etc.), themselves polluting materials when discharged from the plant into receiving waters.

The cooling water discharged into the estuarine receiving waters can also lead to environmental changes given that it is warmer than ambient and may produce an obvious thermal plume covering many square metres. There can be an influx of non-native species and the physiological disruption and acclimation of native species. The temperature changes can disrupt reproductive cycles, for example, summer spawning invertebrates will have their spawning period extended whereas winter spawners will have theirs curtailed. These could lead to population and community changes. The high flows of discharged cooling waters can lead to bed erosion and substratum disruption as well as the dispersion of chlorine residuals and organohalogens into receiving waters. In turn, there will be potential bioaccumulation and cellular and physiological effects of the latter.

Coastal and estuarine power plants have other emitted and produced materials such as aerial emissions, leading to atmospheric inputs, of CO_2, NO_x, SO_x, etc. (greenhouse gases and nutrient inputs). Radioactive materials may be discharged in water emissions as well as fuel waste storage and reprocessing from nuclear stations.

Given the potential effects then it is necessary to produce mitigation measures for the reduction of the effects. There should be plant design and operation to environmental standards (e.g. ISO 14001) and the rehabilitation of areas after use. Trash could be returned or impingement minimized through intake design and deterrent methods or the plant operated on a regime related to fish migration patterns. The discharge apparatus could be designed to reduce outflow scouring, for example, by using "plastic seaweed" on the bed, and to disperse heated water. Heat treatment could be used instead of chemical antifouling and fuel gas desulfurization (FGD) used to control atmospheric inputs.

6.11 Estuarine barriers and barrages

Many estuaries have long had small physical barriers at their upper regions, for example, weirs that mark the head of the estuary. In the last few decades, however, larger permanent barrages and temporary barriers have been placed across several estuaries for various reasons—to protect low lying areas from flooding due to storm surges and high tidal conditions, in order to increase the amenity value of adjoining areas, and for tidal power generation. The interference of hydrographic patterns within the estuary by the presence of such structures will in turn change the physical nature of the estuary and thus its biological structure and functioning.

Low-lying areas are exposed to high tides and especially storm surges, particularly in cases where the tidal wave may get amplified by a constricting topography such as within a funnel-shaped estuary or a funnel-shaped semi-enclosed sea. For example, climatic conditions, wind direction, and atmospheric pressure differences, in January 1953 caused a storm surge to develop in the northern North Sea and then progress down the eastern UK coast and then move around to the eastern North Sea coasts in Belgium and the Netherlands. This produced high tidal levels in the adjacent estuaries and led to overtopping of seawalls. As a result of this, many lives were lost in the United Kingdom and the Low Countries thus leading to the creation of the Dutch Delta Plan and the need for higher seawalls and estuarine barrages in Eastern England.

The estuary of the River Hull, leading into the Humber estuary, Eastern England, has a movable barrier, which is lowered into place during high tides in order to protect the city of Kingston Upon Hull (Fig. 6.25). Similarly, the Thames Barrier (Fig. 6.26) is raised following the threat of high tides in order to protect low lying areas in London. As a much larger structure, the Oosterschelde (Eastern Scheldt) estuary in the southern Netherlands has a storm surge barrier to protect the inner, low lying areas against flooding. Although this is a semi-open structure in which gates are lowered between piers following the threat of high tides, this has changed the hydrographic nature and salinity intrusion of the estuary.

Figure 6.25 Hull barrier.

Figure 6.26 Thames barrier.

Permanent barriers have been built across Cardiff Bay, South Wales, and the Tees Estuary and Wansbeck estuary, North-eastern England, in order to increase the amenity value of those areas. The closure of Cardiff Bay has raised water levels, covered extensive mudflats permanently, and created waterfront areas more desirable for housing. The Wansbeck barrage was designed to create a lake, in which the sea-water exchange is only at highest tides or, during the winter, when the barrage is partially opened. The lake was designed to increase a water sports area although siltation since the barrier was built has reduced water depth sufficiently to restrict such an amenity use. The Tees barrage was opened in 1995 in order to create a high amenity urban area.

The use of tidal power in macro-tidal estuaries has long been cited as an example of renewable, environment friendly, and alternative energy to conventional sources, especially now with concerns about global warming and the build up of greenhouse gases. Electricite De France built the permanent barrier across the Rance Estuary, Brittany, in 1972 for tidal power

generation. A similar and larger structure was also built at Annapolis Royale in the Bay of Fundy, Canada, thus using high tidal ranges in both areas. There are proposals for more major stations in the future. China and the USSR also have small stations in operation, while they prepare for larger schemes and in Britain many estuaries have been assessed with regard to their energy generating potential. In the 1980s and 1990s, many UK estuaries with a tidal range in excess of 5 m were considered and there were detailed proposals and Environmental Impact Assessments (EIA) carried out for tidal energy barrages across the Mersey, Duddon, and Severn estuaries. It was suggested that the Severn estuary, with a 16-m tidal range, would be able to supply up to 20% of Britain's electricity requirements.

An energy-generating barrier would be placed in the lower reaches of the estuary and then it works by allowing water to flow into the estuary and thus creating a head-pond upstream. The barrier can then be closed as the tide ebbs seawards thus creating a head pressure, which can be released as water flows

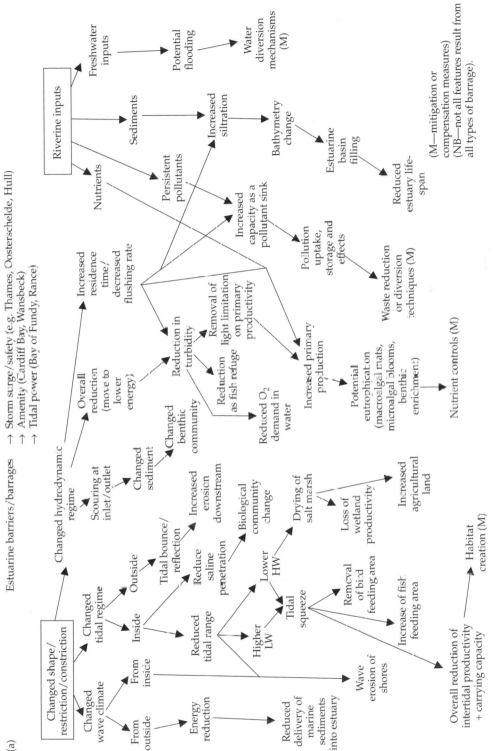

(a)

Estuarine barriers/barrages → Storm surge/safety (e.g. Thames, Oosterschelde, Hull)
→ Amenity (Cardiff Bay, Wansbeck)
→ Tidal power (Bay of Fundy, Rance)

(M—mitigation or compensation measures)
(NB—not all features result from all types of barrage).

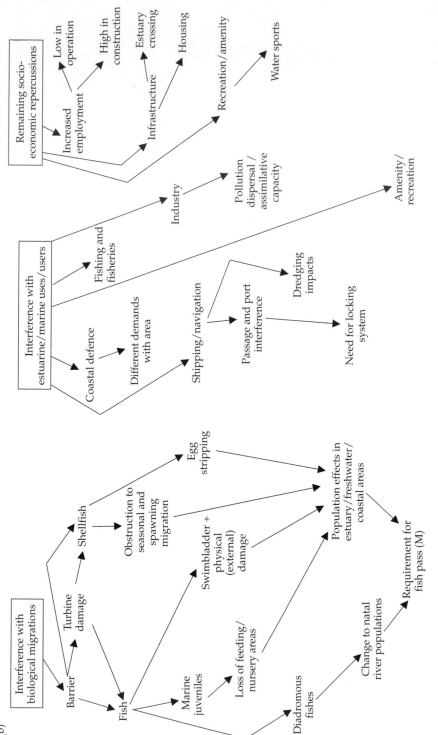

Figure 6.27 Estuarine barrages and barriers—conceptual model of impacts and consequences: (a) changed shape and riverine inputs, (b) interface with migrations, uses and users and socioeconomic repercussions.

back through the barrage. The water movement can turn turbines either just on the ebb or both on the ebb and flood. The presence of a barrier across the mouth of an estuary has the potential to create many changes and adverse impacts in the estuary, the adjacent coastline, and the areas upstream and inland. The main impact is the change to the shape of the estuary and thus the distortion of the hydrographic regime. The changed wave climate will prevent sediment movement from the seaward side although the barrage will retain river-borne sediments that enter the estuary. It may increase wave-induced erosion inside the estuary where the water levels are kept higher than previously. Some of the tidal energy does not enter and so results in 'tidal bounce' which can increase erosion along the coast (Fig. 6.27).

As shown in earlier chapters, fish and shellfish migrations into estuaries are an important and central biological feature. Any barrier will interfere with fish and shellfish migration, both by preventing migration into the estuary and also by causing damage. Pressure changes as organisms move through the barrage openings and turbines will cause external damage such as scale stripping, eye hemorrhaging, and internal damage to swimbladders in round fish, especially pelagic species such as clupeids. Eggs carried on the underside of shrimps and crabs will be stripped by the pressure changes.

The restriction of the tidal wave both entering and leaving the estuary will change tidal regime by reducing the tidal amplitude in that high water is not as high, whereas low water is not as low as previously. This will keep low intertidal areas covered longer, thus restricting feeding by wading birds (but allowing fish longer to feed), and cause upper areas to be uncovered for longer perhaps causing drying out of salt marshes and reedbeds. Hence it is likely that the carrying capacity of the intertidal areas will be adversely affected although this has not been fully quantified.

The barrage is likely to change the sedimentary and water quality regime in the estuary.

Most notably, the estuary will become lower energy in nature and thus suspended material will have greater opportunity to settle out and thus improve the water transparency. This could remove the light limiting control on primary production within the hypernutrified areas, which in turn could increase primary production. Unless nutrient inputs into the estuary are reduced this would increase the possibility of symptoms of eutrophication occurring. The increased residence time and decreased flushing rate will affect water quality through the greater retention of any persistent pollutants from the catchment, which previously would be eventually exported from the estuary.

In addition to the effects on the natural structure and functioning of the estuary, there is interference with estuarine uses and users, for fishing both inside and outside the estuary, and for navigation and shipping. Industry, agricultural, and urban areas upstream will have to improve their pollution control mechanisms and either prevent pollutants from entering the system or divert any discharges around the barrage. Despite this, there may be added benefits for recreation and amenity, with the creation of water sports areas as shown on the Rance estuary, and for infrastructure and communications when a bridge crossing is incorporated into the barrage.

Several of the above adverse effects can be mitigated against although the concern about remaining impacts may prevent other barrages being built. Despite this, the Rance tidal power generating barrage has proved successful and the estuary, while changing from its previous state, is considered to have reached a new equilibrium.

6.12 Other uses

Salt extraction has long been practised on estuarine areas, especially lagoons in warmer countries. For example, the Ria de Aveiro, the major northern Portuguese lagoon, has long supported salt extraction in pans thus providing the salt to produce salted cod, the classic Portuguese dish *bacalau*. Compartments built in shallow areas of the lagoon are allowed to fill with water, which then is allowed to evaporate, and the salt collected. Hence hypersaline waters are produced, which changes the estuarine fauna from brackish tolerant to high salinity tolerance.

Mention should be made of other positive uses of estuaries, such as for recreation or conservation. However, where these require human intervention to improve areas for these then there may be adverse effects. Recreational interests in estuaries range widely; the sheltered waters are used for sailing, swimming, or wind-surfing, often accompanied by the construction of specialist harbors or marinas. Again the construction of any structure will change the hydrographic patterns, the sedimentary regime, and thus the biota. An excessive number of boats will add sewage and antifouling paint residues to the waters, although the most polluting antifouling paint has been banned on small (<25 m) boats.

Many people are attracted to the tranquillity of estuaries for bird-watching or simply escaping from their fellow men. Ornithologists have in particular become a powerful pressure group, campaigning for the conservation of the remaining parts of estuaries. In many cases, however, the presence of large numbers of birds within the estuary may be the result of anthropogenic organic matter, which has displaced natural organic inputs. In Chapter 8 we will discuss the ways in which estuarine conservation may be managed.

6.13 Conflicting uses

We can thus see that the uses of estuaries by mankind are many and varied. In almost every estuary a variety of uses may be occurring, and many of the uses described are not mutually exclusive. Obviously some conflicts do arise between the different users of the estuary. The requirements of fishermen or conservationists for clean undisturbed waters may conflict with an industry or city council looking for the cheapest disposal site for waste material. Despite the many uses of an estuary, for most people living in cities their local estuary is the closest natural ecosystem that they can visit or observe, since the resilience of estuarine fauna has meant that it has survived, whereas other natural habitats, such as the countryside in all developed countries, have been irrevocably altered.

In this chapter we have seen the many ways in which mankind uses and abuses the estuaries of the world. In the next chapter we will consider how we can measure the effects of those activities and then in Chapter 8 we can explore the management of the estuaries.

CHAPTER 7

Methods for studying human-induced changes in estuaries

7.1 Introduction

In the assessment of biological change due to man's activities one needs to ask: What is normal situation and what are the limits of expected variability? Has there been a change from the normal situation and if so can that change be quantified and statistically tested? Is the degree of change significant and can it be related to one particular stress or general environmental perturbations?

An estuarine ecosystem health assessment (or monitoring) program requires an analysis of the main processes of the ecosystem and the identification of known or potential stresses. In order to be scientifically valid it requires the development of hypotheses about how those stresses may affect the ecosystem, followed by the identification of measures of environmental quality and ecosystem health needed to test the hypotheses.

The main aim in any assessment of human impacts is to detect *stress*, which can be defined as "the cumulative quantifiable result of adverse environmental conditions or factors as an alteration in the state of an individual (or population or community) which renders it less fit for survival." Such a definition encompasses techniques at all levels of biological organization from the cell to the ecosystem. Following this there is the need to describe the impacts, to predict where possible the changes under certain circumstances and then, through management actions, to reduce or prevent problems occurring.

This assessment requires a large amount of information:

- What is/are the behavior/characteristics of the system?
- What is the physical/chemical nature of the system?
- What is the physical and chemical behavior of materials added to the system?
- What is/are the behavior/characteristics of an activity in environment?
- Which habitat(s) is/are at risk from being modified or from having materials added to the system?
- Does the addition of materials have an inert or biologically effective action?
- What are the biotic and non-biotic component(s) at risk from the activity or additives?
- What is the behavior of contaminants within organisms?
- What is the structure of the biological system, at all levels of organization, and how does it function?

When an effect is determined this can be regarded as a "*signal*" of response, whereas variability in the features of the system and in any response under noncontrolled experimentation or survey is regarded as "*noise*." Hence we can say that techniques are required to maximize the "*signal to noise*" ratio, that is, to detect a clear change in the biological system against a background of variability. Scientists may then be required to predict and quantify the effects, such that, for example, an environmental manager may ask "what will be the effect of discharging a certain type of waste into an estuary?" or "what will be the effect of building a barrage across an estuary and how big will that effect be?"

All organisms respond to environmental change from whatever cause by a series of responses ranging from subtle metabolic adjustments to dramatic changes such as escape or death. The latter may be the consequence of the stressor not being removed or the organism does not have an ability to cope with it. Figure 7.1 indicates the time-related sequence of the effects of environmental stress on marine and estuarine fauna. The first event of importance to the organism is the detection of change by the sensory receptors, followed by metabolic adjustments and/or behavioral reactions. Mobile animals can swim away from the affected area, animals within the sediments may burrow deeper, and stationary animals such as mussels may simply shut their valves to wait for improved conditions. Depending on the animal and the cause of the stress, the animal may recover its metabolic functions, showing acclimation, or individual animals may be genetically selected to survive the stress, showing adaptation. Alternatively the stress may cause death or serious impairment of normal functions, such as growth or reproduction, leading to changes in populations and communities. Such structural and functional responses can be detected by biological sampling of the components of the estuarine ecosystem.

Because of this sequence and set of responses, we can regard any effects of human activities as acting at one or more levels of biological organization (Table 7.1) in which a change at one level may, if the stressor is not stopped or the biological system cannot cope with the effect, then go on to affect a higher level. Proceeding through these levels, the time of response to environmental change decreases and the inherent variability in the system increases. For example, an organism will respond rapidly to a change in water quality whereas the community will take longer to show changes; a cell is much less variable in its functioning than a community. These features have to be considered when designing any sampling and survey program

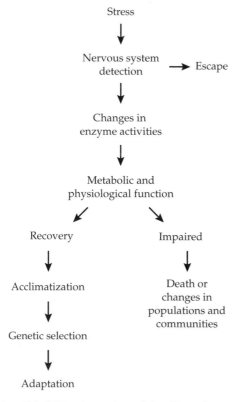

Figure 7.1 Schematic sequence of the effects of environmental stress on estuarine animals. (After Blackstock 1984.)

Table 7.1 Effects at differing levels of biological organization

---

1. Biochemical/subcellular level
2. Physiological effect
3. Change in the individual
4. Changes in the population
5. Community structural change
6. Community functional change
7. Whole ecosystem effects

The complexity (and thus variability) of the system increases from 1 to 7, whereas the time for response (and thus time for detection of effect) increases from 7 to 1.

and when interpreting the data. An effect of a pollutant in a cell will eventually affect the whole organism if the cell cannot deal successfully with the pollutant. Hence a suite of techniques are required to produce data to indicate and measure change at each biological level although those data are only useful once converted into information.

Information regarding changed estuaries is required to answer focused questions:

• Do the indigenous populations reflect the hydrophysical characteristics of the estuary?
• Do the populations and growth rates differ from similar areas?
• What is the trophic system and how do any organisms link with the general food web of estuaries and the adjacent coastal areas?
• What is the structure of the communities and how do their populations function?
• How have the health of the individuals or the populations been affected by man's activities?
• Have any environmental or ecological quality standards and objectives set for the estuary been met?

A wide variety of techniques has been developed to study the changes, which occur in estuarine organisms in response to human activities, including pollution, ranging from laboratory-based toxicity testing procedures, through to field-based population sampling programs, and including physiological, cytological, biochemical, genetic, behavioral, pathological methods as well as bioassays and ecologically based methods.

There are many techniques suitable for detecting change, each of which is more suitable than others for particular stressors, for example, some will detect the effect of heavy metals from a polluting discharge, whereas others may detect community change through sediment changes caused by creating an obstruction in an estuary. The techniques should show a consistent relation of the response to individual or species pathology and it should be known whether the response is to either specific or general stressors. The method should be precise in indicating change and have a large response compared with the variability (i.e. a high "signal to noise ratio") but in order to produce a cost-effective technique, which can be employed in management, there should be a rapid response after contact with the stressor. In the case of pollution, the test should be applicable to a wide range of phyla and their life histories should demonstrate a dose–response relationship. Some techniques are suitable for use for a general survey, whereas others are a diagnostic test to give a detailed investigation in a "hotspot." Finally, the method should be relatively straightforward and at a reasonable cost. Despite the many techniques being developed for assessing change, it is, however, necessary to be cautious as all methods have problems, many are in early stages of validation and are not necessarily suitable as working techniques. In many cases, the sources of variability in the method have not been determined or quantified accurately.

The science of impact assessment has been complicated further by the use of interchangeable terms: the studies can be separated into surveillance and condition monitoring, compliance monitoring, and diagnostic testing, although different management protocols may use different terms for these (see also Chapter 8). *Surveillance*, sometimes called *condition monitoring*, is the assessment of a small or large area, at one time or over time. In contrast, *compliance monitoring* assesses those characteristics but with a given end point in mind—that is, what is the expected or required status of an area, its populations or communities—in essence whether the area complies with what has been decided as the preferred status. Condition and compliance monitoring may also be termed *operational monitoring* and, if the results are used to check whether any legal licence has been met, also *Statutory monitoring*. *Diagnostic testing*, also termed applied research, a trigger-exceedence study or *investigative monitoring*, can then be used to determine the cause of change.

7.2 Individual change

7.2.1 Toxicity testing

The simplest response of an individual organism to a stressor such as a pollutant is through toxicity testing, which involves placing organisms in static, or preferably flow-through tanks, with known concentrations of the pollutant. The results can be expressed in different ways, but all follow the basic concept shown in Fig. 7.2, where response is related to dose. For a non-essential element, small doses produce little response, but as the dose increases the response increases. Below a certain threshold the responses are considered as acceptable changes, but above that level the responses are considered unacceptable. For an essential element, for example, copper, small doses are vital for life, and a lack of the small dose, as well as large doses, produce unacceptable change. In between these dosages is an ideal dose. All toxicity testing is designed to determine the safe threshold between acceptable and unacceptable change.

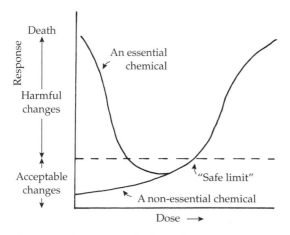

Figure 7.2 The response of a "typical" population to changes in the dose of essential and nonessential chemicals. The position of the dotted line on the arbitrary vertical axis varies with the toxicity of the chemical, and for some poisons, e.g. carcinogens, the dotted line may not exist, there being a continuous gradient of response with concentration. The intercept of the dotted line with the response curve indicates the "safe" dose limit. (After Pascoe 1983.)

Acute toxicity testing measures the median lethal time, LT_{50}, that is the time at which half the population dies at a given dosage. The same test can be used to derive the median lethal concentration, LC_{50}, for a specified time, conventionally 96 h. Since such tests only determine the lethal (or acute) stress of a pollutant, it is common to multiply the results by an application factor (0.1 or 0.01) to determine a "safe" level for the pollutant. For example, if 50 ppm of a substance is required to have a lethal effect then a regulator will only allow a company to discharge 5 ppm or even 0.5 ppm. It has to be remembered, however, that pollutants are not discharged in isolation. A complex discharge from an estuarine petro-chemical plant will contain hundreds if not thousands of chemicals, each of which may interact before, during, or after discharge and may have their chemistry modified by the nature of the receiving waters, such as its temperature or its rapidly changing salinity. Because of this, Direct Toxicity Assessment (DTA) (also called whole-effluent testing) is increasingly being used to detect any synergistic or antagonistic effects of the pollutants in combination.

Lethal tests are widely criticized, and many investigators recommend sublethal (or chronic) testing of pollutants as an alternative. Chronic testing may, for example, look at behavioral changes, reproductive success, or metabolic responses of animals subjected to low doses of a pollutant. The use of low-level, long-term testing is essential if we are to fully understand the problems caused by dilute concentrations of pollutants, however, they are more difficult to administer, control, and reproduce than acute tests. Thus, despite the drawbacks, acute testing remains the principal method for the laboratory assessment of pollutants, which lead to guidelines for the disposal of toxic materials.

Although estuarine animals are often used, for example *Corophium*, *Nereis*, and *Macoma* have all been greatly used in toxicity testing, there is an increasing use of macroalgae in toxicity tests, both in the laboratory and in the field. For example, mesh bags containing excised discs of the green alga *Enteromorpha* or the growing tips of the brown alga *Fucus vesiculosus* can be located in effluent concentrations and then the viability of the cells or the growth of the tissue can be measured as a response.

7.2.2 Behavioral bioassays and sediment testing

In the past, many toxicity tests have used organisms placed in water. As shown in the previous chapter, especially in sediment-rich areas such as estuaries, very often pollutants enter sediments and may get bound into them. The behavior and chemistry of pollutants are likely to be changed by the sediments' physical and chemical structure. Because of this, sediment bioassays are increasingly used— these include placing the lugworm *Arenicola*, the bivalve *Macoma*, or the amphipod *Corophium* in sediments and determining its survival or behavior. The latter may include its ability to burrow, the speed of burrowing, and in the case of the lugworm, its rate of faecal cast production. These are often termed behavioral bioassays and may be used as a more realistic, chronic, and perhaps early-warning indication of change.

The main aim is the measurement of deviation from normality in any aspect monitored. In the development of behavioral bioassays, this also indicates the problems with such assays. There is a difficulty in determining normal behavioral patterns and quantifying the deviation (signal) due to a stressor. For example, in the burrowing bioassay using the estuarine bivalve *Macoma*, there is the need to determine how it normally responds to sediments and how that response differs when the sediment is polluted. Another example of behavioral bioassays used in estuaries is the KEMA Mussel Monitor, a chamber supporting bivalves such as mussels and cockles in which their valve movements are measured in relation to water quality—this has been deployed

in power station and near chemical industry outfalls. Finally, regulators, especially in North America, have made large use of a bioassay Microtox® using luminescent bacteria in which a measure of the decrease in their light output gives an indication of the toxic nature of any wastewater.

7.2.3 Toxins and bioaccumulation

When studying stress it is necessary to study not only the biological system but also the physical and chemical system. For example, pollutant uptake may be directly from water, across the gills of fish, from food during its passage through the alimentary system and, in the case of bottom-dwelling organisms (e.g. flatfish) across the skin after exposure to residues in sediment. The relative importance of these routes differs between organisms and chemicals and depends on factors such as salinity, temperature, and pH of the water, the presence of other chemicals/metals in solution, the size and feeding state of the animal, and the chemical form and nature of binding of the pollutant. Hence the monitoring program needs to consider all of these aspects if any changes are to be correctly and fully interpreted.

Once taken up into estuarine organisms, some pollutants may induce a detoxification mechanism, which protects the organism against the harmful effects of the chemical, whereas others do not. It is of benefit to the cell to reduce the availability of free pollutants in the cell, and thus to limit or prevent interactions with cellular components, which may be harmful to the organism. For example, organic pollutants may be metabolized by inducing enzymes such as the mixed function monooxygenase system (induced through the biochemical pathway cytochrome p450 system), which can then produce low-toxic hydrophilic (water-affinity) metabolites, which can be rapidly excreted. Occasionally the production of active intermediate metabolites (such as carcinogens) may cause more damage to the cell than the original compounds (see

Lawrence and Hemingway 2003 for further details). The release of stress proteins may repair some of the damage caused by pollutants.

Binding (sequestration) to another molecule also reduces the bioavailability of pollutants, which can lead to either excretion or storage. For example, Heavy Metal Binding Proteins, a form of metallothionein proteins, which can bind metal ions, become induced when there is exposure to high metal concentrations. The presence of these detoxification mechanisms or the bioaccumulation of the pollutant will indicate recent exposure. Accumulated metals may be stored in specific organs and slowly excreted. In many vertebrates, including fish, the main storage organ is the liver, which can therefore be expected to give a better indication of recent contamination, whereas muscle tissue may give long-term uptake and storage (Fig. 7.3).

At the whole-organism level, pollution may act directly by increasing mortality, or indirectly by causing changes in behavior such as impairing/reducing an organism's efficiency of foraging ability thus reducing food uptake and thus somatic production. Hence, in turn, these effects on individuals can reduce population growth, or produce population decline. All of these mechanisms at cellular and individual levels use energy and resources, which are therefore not available for production thus creating a population effect. In turn population responses will produce effects at the community and ecosystem levels, hence the need for monitoring exercises to link the levels of biological organization. This whole field of biomarkers, used to indicate responses to pollutants, is rapidly expanding and gives great scope for detecting an early-warning of pollution effects.

Bioaccumulation studies can additionally be conducted to indicate both sublethal stresses within individuals as well as contamination in estuaries (Phillips and Rainbow 1994). There are many benefits in the use of organisms over other methods for monitoring aquatic contamination, for example, the environmental and management relevance, especially where food species are used, as well as the relative ease of measuring the often higher levels of pollutants in the tissue than in the water column. The degree of bioaccumulation has repercussions for the organism itself, its predators (including man in the case of commercial fishes), and its progeny. Because of this, bioaccumulation studies are widely used by statutory bodies charged with assessing and protecting the estuarine environment.

The assessment of stress and the degree of contamination of estuarine fauna at the subpopulation level require the use of common,

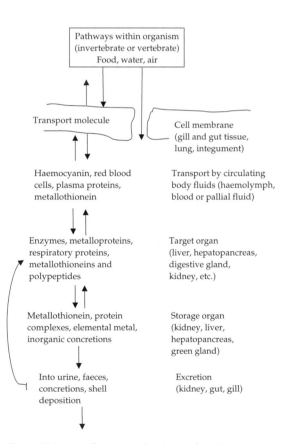

Figure 7.3 Fate of trace metals taken up from the environment and diet within marine organisms. (After Engel and Brouwer 1984.)

representative indicator (or sentinel) organisms that can be used for bioassay or bioaccumulation studies. Bioaccumulation monitoring programs require species to be used, which fulfill several criteria:

- being sessile or sedentary and thus representative of the study area
- abundant in study areas, easy to identify and sample and providing sufficient tissue for analysis
- able to tolerate wide ranges of contaminant concentration and of physico-chemical variables such as salinity, thus even allowing an experimental approach to measure uptake
- provide a simple correlation between the tissue concentration and the ambient levels.

Most importantly, the accumulation strategy of a potential biomonitor for a particular contaminant should be known, for example, whether it will take up pollutants in the particulate form common to highly turbid estuaries, or from solution because of their solubility and lipophyllic nature. Species such as estuarine/marine macroalgae such as *Fucus vesiculosus* and *Enteromorpha*, the suspension-feeding marine mussels *Mytilus*, and the estuarine fishes: the flounder, *Platichthys flesus*, and the eelpout, *Zoarces viviparous* have many of the characteristics listed above necessary for sentinel organisms. Monitoring of the health of estuaries will require the use of a sentinel organism or at least estuarine residents with a small home range, hence reducing the uncertainty in the interpretation of the data and increasing the link between the health of the biota and the ambient stressors. For example, as sediments receive and store many pollutants then organisms feeding on these or on their infauna are vulnerable, for example, the flounder *P. flesus*, or wading birds (Fig. 7.4).

7.2.4 Individual health and external and internal body condition

It has long been acknowledged that younger organisms are more susceptible to the effects of pollutants, hence a larva is more susceptible than an adult and an embryo is more susceptible than a larva. Their small size and often unprotected surface, which thus gives them a large surface area to volume ratio and thus a large surface area in contact with any

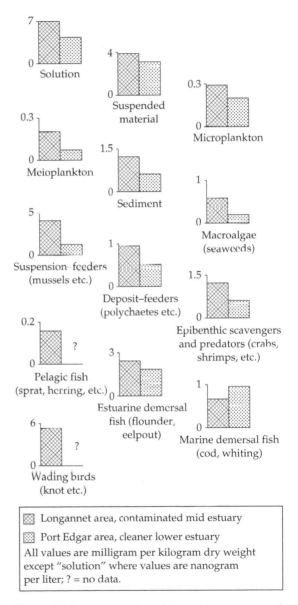

Figure 7.4 The concentrations of mercury in components of the Forth Estuary, Scotland. (Data from Elliott and Griffiths 1986, figure from Elliott and Hemingway 2002.)

waterborne pollutants, their place in the upper water layers while planktonic and their high metabolic rate, which requires the rapid movement of materials into the organism, all increase the uptake of pollutants. This has led to the production of developmental bioassays such as the oyster embryo bioassays, which measures the rate of development and the success of development in reaching the D-shape characteristic of most bivalve larvae.

The health of individuals and the effects of pollutants discharged into estuaries may be shown by the presence of morphological changes. These have been used widely and are potentially valuable in pollution monitoring programs and especially with high-profile estuarine organisms such as fishes (Elliott and Hemingway 2002). Within fishes, for example, morphological disorders range from mild skin discolorations to major skeletal deformities such as blemishes, lesions (raised areas of skin, which may or may not be broken or infected), lymphocystis-type nodules (generally found on the dorsal, ventral, and anal fins and the caudal peduncle), fin rot, eye-deformity, mouth ulceration, and skeletal deformities such as scoliosis (spine curvature). Because of this, it is valuable to record routinely morphological anomalies during surveys of fish communities. These give a simple qualitative analysis of the response to pollutants and other stressors within the area, and they may alter the structure and functioning of the populations although the effects are difficult to quantify. There may be changes in the reproductive capacity, or success, of the organisms, that is, fewer number of eggs may be produced, or fatal deformities may arise within the larvae as a result of contaminants within the environment. There will, however, be variations in their incidence due to season, age, migratory patterns, and species as well as that due to external stressors. The latter may be anthropogenic, for example, pollution and contamination, but also abnormal levels of natural features such as salinity and temperature. As such, it may be difficult to determine cause and effect relationships with respect to fish health. In addition, it may be difficult to determine whether an observed abnormality is due to a stressor acting on an individual or on its previous generations and, especially in estuaries and for mobile species such as fish, whether it is operating in the area where the animal was caught or elsewhere. This is particularly important with migratory fishes that use the sea and/or freshwaters as well as the estuary.

The flesh condition of an estuarine organism, an index of fatness, and the relationship between weight and size or volume, and its overall individual health may also provide valuable information on the health of the system. The condition factor varies within a population, especially between life stages, but also within feeding status and thus it reflects emaciation created by spawning, poor food conditions, or overwintering. Thus any natural or anthropogenic stressors which affect the energy balance will affect condition, for example, the technique of assessing scope-for-growth.

7.2.5 Molecular and genetic techniques

Molecular techniques are also included in the suite of approaches available for determining change in the estuarine system—for use for stock discrimination, that is, differentiating between populations, and for assessing stress damage. The chronic or occasional presence of any substance in the environment can exert a selective pressure upon organisms and the pollutants may directly damage the genetic material thus producing mutations (Beardmore *et al.* 1980) (Fig. 7.5). In addition, within estuarine environments, genetic changes to populations due to escapees from aquaculture are also possible. While there is the need to determine the responses (as effects) of populations in adapting (or attempting to adapt) genetically to stresses caused by human activities, this has not been studied in estuaries. Similarly, mutational changes have not been studied as the result of estuarine stress.

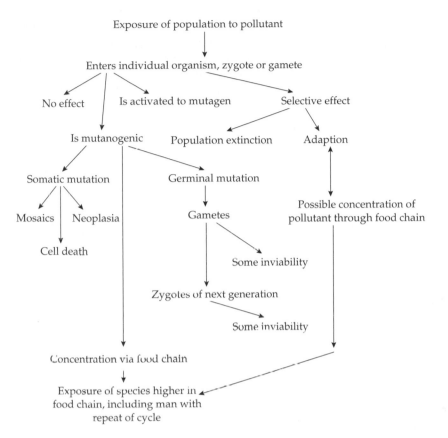

Figure 7.5 Generalized scheme of the genetic effects of pollutants. (Reproduced with permission from Beardmore *et al.* 1980.)

The speed of response of the population is related to the intensity of the selection pressure applied and by the ability of the population to respond to it. This in turn is influenced by internal factors such as genome size, number of chromosomes, amount of recombination, population size, and generation time. Beardmore *et al.* (1980) suggested that the following methods of response to polluted conditions are possible:

- an organism may react entirely phenotypically, usually in its growth rate or by modification of its reproductive output
- by an increase of one or more genotypes or sibling species at the expense of others.

These aspects have not been specifically studied in estuaries, although estuarine resident species such as the common mussel *Mytilus*

edulis and the harpacticoid copepod *Tisbe* have been found to have a complex genetic population structure, which may result in their tolerance to environmental stressors found in estuaries. Other genetic assessments are now being carried out, for example, to determine the effects of escapees from the culture of salmonids in fjordic systems and imported oysters in estuaries, particularly as these will have been reared especially for culture (Gosling 2003). In order to produce a reliable monitoring technique, the main needs are:

- to investigate populations exposed to pollutants to establish the effects on fitness, congenital anomalies, resistance to disease, lifespan, fertility, etc.
- to determine the variation, in time and space, of the gene pools of populations living in clean and in polluted environments, and

under laboratory conditions; to analyze the genetic basis of any observed adaptive shifts in phenotype across a range of species

- to assess the occurrence and distribution of mutagens in the marine and estuarine environment
- to test both singly and in combinations the effects of these mutagens on a representative range of species
- to establish a monitoring scheme to screen appropriate species for mutation and allele frequency changes.

7.3 Population responses: Population dynamics, growth, mortality rates, and population models

If there are sufficient changes to individuals of a given species and these are not compensated for by the organisms then they will lead to population changes. An analysis of the population dynamics, that is, the relative sizes of age stages in the population, will indicate if younger and more vulnerable stages have been adversely affected. There is the need to measure the effects of activities on predator–prey relationships and oscillations, especially in relation to birds and fishes using the estuary. For example, if pollution or other stress within an estuary removes a preferred prey size then this will affect a predator.

Growth and mortality of individuals and their effects on population dynamics are an important measure of determining whether an estuarine area is functioning as is expected given its environmental characteristics. The data can then be used within population models aimed at predicting the yield. These techniques have been most widely used on fish populations. The population structure of a species can be described after measuring a representative set of organisms and then plotting a size–frequency histogram. The presence of the modes within that histogram will indicate the average sizes of cohorts and then changes in the frequency and average sizes on successive sampling occasions will indicate the mortality (survival) and growth

of the cohorts. Growth can be determined by several methods—by marking organisms and releasing them before re-capture and re-measurement, such as used extensively with fishes, especially in freshwaters, but also with sedentary invertebrates such as estuarine and marine bivalves. The use of growth ring discontinuities in scales or otoliths of fishes and shells of bivalve molluscs can be used to determine age at each size. Similarly, growth can be estimated either by analysis of length–frequency histograms or from the mean size of an organism of a known age at the time of sampling. If one or more particular size groups (cohorts) can be distinguished in a length–frequency histogram of a sample and followed through successive samples, the modal sizes plotted against time will produce a good estimate of growth (for further details see Crisp 1984). Alternatively, if all individuals of a species in a sample can be aged, the mean length of each age group can be used to estimate annual growth or growth between successive samples over a shorter time span.

Once growth as size or weight at each given age has been determined then it can be summarized in numerical models, which may be theoretical or empirical—the former are designed to describe the sigmoid growth curve followed by most organisms whereas the empirical models merely give a best fit line to the observed growth at any time or place (see Bagenal 1978; Elliott and Hemingway 2002 for further details). While growth may be reduced in estuaries as the result of pollution, it may also be reduced in the case of primarily marine species such as mussels due to the reduced salinity. The reduction in population size, that is total mortality, is necessary to determine the loss of organisms through natural or anthropogenic stressors. It can be expressed numerically in two different ways— the actual mortality rate (A), as the fraction of organisms present at the start of a period of time which die during that period and where the survival rate s is $(1 - A)$, and as the instantaneous mortality rate (Z), an exponential rate, of reduction in numbers (see Gee 1983).

7.3.1 Biological production

Biological production is a valuable parameter and will indicate the nature of the area, the success of an area supporting a population, the success of the populations themselves, and the material available for higher and detrital foodwebs. For example, where there is the need to determine the carrying capacity of an intertidal estuarine mudflat for wading birds then it is important to know the amount of food produced. The production may be determined by species (individual, cohort, and population) or by communities (by adding the production by all the age groups of each species in the estuarine assemblage) (see Crisp 1984 for further details). For an individual, the "physiological" production (P) depends on the food consumption and the metabolism of the species and equals the sum of P_g (tissue growth), P_r (production of genital products), P_s (secretions such as mucus), and P_e (dead tissues).

The population production available for predators (including Man) is termed yield and can be determined using the cohort production based on the growth as size or weight of individuals belonging to the same age–class (the latter is termed the Allen curve). This indicates the relative decrease in abundance of a cohort, through natural mortality together with the increase in average size, and thus weight, of the cohort members. Production may also be estimated using growth and average biomass data where the latter is the organic matter contained in the tissues, often measured as ash-free dry weight or measured as wet weight and then calculated using conversion factors.

Production values are most accurate when based on a large sample size which, given the nature of estuarine populations, will be of a few dominant species. Production can be estimated using $P:\bar{B}$ ratios, derived elsewhere or from the literature, together with an average biomass from the area under study (Elliott and Taylor 1989). A theoretical value of productivity (e.g. $P:\bar{B} = 2.5$) derived from other studies can often be used to calculate the production from the mean biomass data. This may only be an approximation as this ratio depends on the species, the age of individuals, and the environmental conditions; it is often underestimated when juveniles are concerned, although Elliott and Taylor (1989) found this value to be suitable for estuarine invertebrate species.

Although there are difficulties in estimating production, its values may be used with biomass data, to assess the functioning, as the transfer of material during feeding, of an area. Most methods for estimating the production of estuarine populations and communities require input data on growth rate, mortality rate, and abundance at least at one time, although production may be extrapolated for groups of species (trophic levels or communities) from production estimates of other parts of the food web (notably the primary producers).

7.4 Community level assessments

As shown in the previous chapter, any change to the overall nature of the estuary such as hydrographic and sedimentary features, will change communities, especially the bed community. Changes to sediment type or quality and water quality will sooner or later affect bed community structure, hence these remain the mainstays of assessments of estuarine environmental change. All laboratory-based assessments of pollutants should be supported by comprehensive field-based studies of the actual effects of the pollutants on estuarine organisms. For a field assessment of the effects of a pollutant discharged into an estuary, or of the change to the community through engineering works, the principal method is to collect samples of animals or plants from areas or times believed to be affected, and to compare them with samples from areas or times believed to be unaffected. This has given rise to the BACI–PS approach—the Before After

Control Impact–Paired Series statistical approach to determining spatial and temporal change (Schmitt and Osenberg 1996).

Stationary organisms such as the infauna and the macroalgae are usually the mainstay of monitoring programs because their community structure will reflect the prevailing conditions in an area, they are usually easy to collect and their taxonomy is well developed and thus studies have a high degree of quality control. A typical estuarine sampling program will collect samples of the benthos from the intertidal or subtidal areas of an estuary. A survey to provide a general view of the quality of the estuary requires a widespread sampling pattern, which may have a regular grid of samples, or a sampling pattern stratified according to habitat type. In contrast, the need to determine the effects of a point source or of the start of a discharge would have a sampling design centered on the source and then with an increasing distance between the sampling stations with distance from the source. Needless to say, the sampling grid has to reflect the environmental variability of the estuary—the hydrographic and sedimentary regime.

For the determination of community changes, there can be an *in situ* assessment in which organisms are identified on site, this may be to a lower taxonomic level and is termed a Phase-1 or "skilled eye" method. This will involve the recording of visible organisms, such as the epifauna or macroalgae, or the sieving on site followed by the recording of those observed in the sieve—the sieve residue. In contrast, for a more detailed assessment, the material collected would be returned to the laboratory, and all the animals and plants collected identified to the highest taxonomic level-species level where possible. A comparison would then be made between the estuarine community close to the source of pollution and those farther away. Such comparisons can be made from the basis of the number of species collected, the abundance of individual organisms, the biomass of the organisms, or by chemical analysis of the specimens. There are many designs of sampling strategy, each with an ability to answer particular questions and to produce required information (Table 7.2).

Community assessments have to detect and quantify changes to the dominant flora and fauna, for example, the presence and extent of marine mussels producing biogenic reefs or strands of macroalgae which may have extended as a response to increased nutrient retention in the estuary. Both of these may be mapped using rapid surveying techniques or by using aerial photographs. For example, the development of macroalgal mats that are causing concern , as the possible result of increased nutrient runoff, can be mapped using airborne surveillance techniques.

The measurement of any biological change within the estuary has to be explained by linking it to the prevailing environmental factors, both natural and anthropogenic. These may include the natural physical and chemical nature of the sediments and whether they have been changed by human activities. This is especially important in estuarine sedimentary areas. There is the need to know the hydrographic patterns such as tidal, wind driven, density (salinity) currents, and residual currents such as Coriolis. Each of these will indicate the areas liable to be influenced by any discharge, the receiving area and near-field and far-field effects. Similarly, as indicated in Chapter 6, creating structures within an estuary will change current patterns and thus sedimentary regimes. Such knowledge cannot, however, fully explain the status and changes to the biological features—it is also important to measure associated biological parameters such as predator–prey relationships, the availability of food, and faunal and floral associations. A full description of field methods for sampling estuarine organisms is given in Baker and Wolff (1987), Holme and McIntyre (1984), Kramer *et al.* (1994), and Elliott and Hemingway (2002).

Table 7.2 Examples of sampling design to determine estuarine change

Design	Examples
Transect	Down the shore (high, middle, low; tidally standardized profile or fixed points)
Transect	Based on the direction of a hydrographic feature, for example, tidal movement
Skilled eye	For biotope assessment and quantification
Ellipse	Aligned according to the dominant current direction but with latitude for movement perpendicular to that direction
Circular grid	Based on a null hypothesis of no change in any direction around an interest point
Circular grid	Hexagonal variant of central treatment area surrounded by a control area
Box grid	Null hypothesis of no spatial effect but with regular spaced samples (e.g. based on lat./long. points, intersections)
Point sampling	Random sampling points across an area
Point sampling	Stratified random in which a given number of sites are arranged in each habitat type (stratum)
Treatment area vs control	With replication, based on well-defined locations of both areas
Increasing spatial separation	On the basis that the effect with decrease in intensity away from the source, hence a grid or transect with station separation increasing (log, semi-log interval)
Temporal surveys	To cover change with time as well as different impacts over time, for example, for a pipeline to cover preconstruction, postconstruction/preoperation, postoperation, postdecommissioning
Tow	For nekton or benthos (using a dredge, trawl, plankton net) to give geographic coverage (e.g. axial in estuary) over a known distance
Tow	For nekton or demersal fauna (using a dredge, trawl, and plankton net) using tidally standardized tows

7.4.1 Numerical analysis of estuarine data

There are many numerical methods relevant to the sampling and interpretation of estuarine communities, especially invertebrate communities (e.g. see Krebs 1989; Table 7.3) and in particular the estuarine benthos in relation to pollution or other stressors (Elliott 1993, 1994). Univariate methods are used for the interpretation of the primary and derived community parameters such as species richness, abundance, biomass, and diversity of the community. Increasingly multivariate methods are used any data set in which a set of attributes (e.g. species presence or abundance) has been measured across a set of samples, which may be sites, stations, or areas.

The community structure refers to the species composition of an area, which can be described according to the number of species and the amount of biomass present, the diversity of the assemblage, and the distribution of individuals and biomass among those species. Estimates of abundance indicate the size of the populations of each species although their accuracy is dependent on the methods used and on samples being representative of the whole population.

Many features can be used to describe an estuarine community and interpret changes in relation to human impacts, environmental variability and biological processes (see Table 7.3). Community or assemblage analysis can be carried out using the primary community variables of taxon (usually species) richness (S), abundance (A), or biomass (B) data. The different measures can be used at the binary (presence/absence) level (taxon richness) or as quantitative data (abundance and biomass), depending on the reasons for the study and the methods used. Quantitative

Table 7.3 Concepts and techniques for detecting and indicating change in estuarine and marine communities

---

General texts giving an overview of the methods and concepts

Sampling design, ability to detect effects cost-effectively, replication required, use of multiway ANOVA for inter- and intra-site analysis, precision of means derived

Time-series analysis, moving-average data treatment

Concept of indicator species; opportunist transition and climax community species

Determination of numerical or biomass dominants (10 highest ranked species, species composing the first 50% of cumulative abundance or 90% of cumulative biomass)

r-strategists and T-strategists versus K-strategists

SAB trends (primary community parameters, species (or family) richness, abundance biomass—Pearson–Rosenberg Model)

H', J', $1-J'$, E, B/A, A/S (secondary or derived parameters, biomass and abundance ratios)

ABC (abundance biomass comparison) characteristics

$S(N-B)$ and DAP index (resultant difference (and difference in area by percent) in abundance and biomass profiles)

SEP (Shannon–Wiener evenness proportion, as development of ABC and comparison of H' (or E) derived on both abundance and biomass)

Rarefraction, log-normal and k-dominance patterns (graphical representation of community characteristics)

AMOEBA-type changes (abundance of dominants from one period to another)

Change of feeding or eco-trophic guilds (e.g. Word infaunal trophic index, UKITI)

Source: Elliott and Hemingway 2002, references for the topics therein.

data allow a higher level of analysis, with the determination of the rarity of a species. In addition to this taxonomic approach to community structure description, analyses of biomass/size spectra, and functional guilds can be used and are an increasingly valuable means of interpreting community structure. The functional groupings or guilds can be based on the ecological preferences of species, their reproductive strategies, and their feeding modes (Elliott and Hemingway 2002 for details). Other functional aspects can be determined in communities, such as using ratios to compare taxonomic types or organisms with different habits.

The primary–community variables can be used to derive other secondary variables such as diversity (e.g. H' Shannon–Weiner index) and Biomass (B/A) and Abundance ratios (A/B). These univariate data can then be interpreted and interrogated using conventional hypothesis-testing techniques (e.g. ANOVA) although complex community data, such as the abundance of each species in a sample, are often more suited to analysis by multivariate statistical techniques. Multivariate numerical

techniques such as cluster analysis can be used to define communities, that is, as a set of attributes such as species which are occurring together, whereas other methods define continua of both taxa and environmental features, for example ordination analysis. Any area sampled will easily produce a matrix of attributes (e.g. species presence, abundance, and/or biomass) against sample (e.g. site, salinity regime, and area). There are a variety of methods available for multivariate community analysis such as ordination and cluster analyses (Gauch 1982; Begon *et al*. 1990; Jongmann *et al*. 1987; Elliott 1994) and gradient analysis. These allow arranging the different species' abundance along a selected gradient or allow communities, or sites, to be ordered along continua according to similarities in their species composition and abundance. These enable the analysis of community attributes, taxon, abundance, or biomass, at a series of stations or sites, and can be performed as either Q- or R-mode. The former describes similarities between different stations according to the taxa present, its abundance or biomass, while R-mode classifications group taxa according

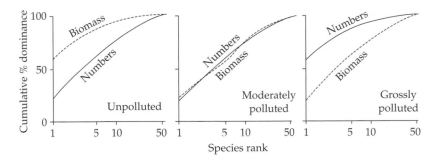

Figure 7.6 Hypothetical *k*-dominance curves for species biomass and number, showing unpolluted, moderately polluted, and grossly polluted conditions. On the horizontal (*x*) axis, the species are ranked in order of importance, using a logarithmic scale, and on the vertical (*y*) axis, the percentage dominance is plotted, using a cumulative percentage scale. (After Warwick 1986.)

to affinities in their distribution. For further information see Elliott and Hemingway (2002).

There are several graphical techniques to indicate change such as the S-A-B curves (Pearson–Rosenberg model) described in Chapter 6. As a further example, Warwick (1986) assessed the pollution status of a benthic community by plotting the distribution of the numbers of individuals among species— the abundance–biomass-comparison (ABC) (Fig. 7.6). In unpolluted conditions the benthic community will be dominated by one or a few large species, each represented by rather few individuals. In polluted situations, benthic communities become dominated numerically by one or a few small species; each represented by many individuals. However, caution has to be exercised as an influx of small organisms during recruitment can also create similar patterns in the community.

There is an increasing amount of software, especially as integrated ecological and biological packages, used to calculate this wide range of numerical techniques and indices. The increase in computing power has produced a greater and easier potential for analyzing large data sets. The development of sampling, the treatment of replicates, the use of power analyses in determining the level of sampling required and the cross-calibration of

numerical treatments all, however, require further effort and testing in estuarine habitats. Trends in all of these features can be determined and used to indicate the functioning of the community in a defined area at a specified time; these can then be compared to data from other times and places (Moss *et al.* 1982). Thus it is possible to determine the "normal" situation from a variety of sites and populations and then show if the study area is different, and if so, whether this is due to natural or anthropogenic factors. Many techniques have been used to separate the effects of natural disturbance/stress (such as found in estuaries and sedimentary unstable areas) from anthropogenic stress and this has resulted in the widespread use of indices of environmental quality. The methods are required to indicate both the human and natural importance of the areas, habitats, and populations.

Community studies, and especially those involving the macrobenthos have long been the mainstay of estuarine impact assessments, often because its features and taxonomy are well known and its response to pollution and stress are well described. The model for change (see Table 6.2) indicates the nature of change and thus the characteristics of the fauna, which should be studied within an impact assessment. These changes underpin especially the ABC and SAB techniques. As mentioned throughout this book, the presence of large numbers of small organisms within a few species, which is characteristic of stressed and polluted conditions, is also typical of natural estuarine populations and so this has to be borne in mind during impact assessments.

In order to assess the effects of disturbance from pollution, dredging or other activities, field experiments can be established in estuaries, whereby an area is deliberately disturbed or defaunated and the recovery of the fauna observed. In such a study in Alewife Cove, Connecticut, USA, Zajac and Whitlach (1982*a,b*) found that the response to disturbance was very variable, and that no set pattern for estuarine re-colonization succession could be detected. Thrush and Roper (1988) have also shown that re-colonization studies are difficult to assess for pollution monitoring.

7.5 Integrated estuarine assessments and ecosystem changes

While some techniques are used in isolation, it is now becoming more common to carry out integrated environmental assessments in which a combination of complementary techniques is used. For example, the sediment quality triad approach (Long and Chapman 1985) combines three measures: of community health, as seabed quality as shown by benthic community analyses, of individual health as shown by an invertebrate bioassay such as the suitability of sediments for burrowing by infaunal species such as *Macoma*, and lastly an environmental (abiotic component) such as the degree of contamination (e.g. the concentration of persistent chemicals in sediments). The latter gives a "cause" component to help to explain the former two "effect" components.

There are many methods at our disposal for detecting change in estuaries. All methods for detecting the effects of pollutants and other causes of change in estuaries are a compromise between the desirable, the practicable, the affordable, and the reliable. It is generally true that almost any method will be able

to detect severe sources of pollution, or "hotspots," but more subtle change requires a wide variety of approaches, and the skill and ingenuity of the scientist may be severely tested. Such skill is especially needed for estuarine studies, since a wide range of natural environmental variables, such as salinity, temperature, or sediment type, occur alongside any perturbations induced by mankind's use and abuse of the estuary.

In managing estuaries, we are often interested in their ability to maintain the high predator populations, especially the fishes and birds, and so we are interested in knowing the trophic structure of the area. For example, we need to know if land-claim removes an area supporting large populations of primary consumers or if sediments change sufficiently such that an area no longer supports deposit-feeding infauna. Hence the loss of an estuarine area through land-claim or a change in its tidal height or sedimentary nature through accretion, will influence its foodweb. Because of this, analyses of stomach contents and feeding behavior will determine the use of areas and the repercussions of loss of areas. Those analyses can give prey preferences, prey importance, and the amount of prey required to support the population. They can also give niche breadth, for example, using diversity indices to show what number of prey items are taken and the opportunistic or specialized nature of predator feeding, and niche overlap, that is, the similarity of feeding between two predators either within or between species to indicate possible competition.

We can define the ability of an area, both according to the amount of space and food available, to support those higher predators as its "carrying capacity." Hence any change in that carrying capacity, especially because of a loss of an estuarine area, a habitat or a biotope, is of major concern to the estuarine managers. A unit of the ecosystem, which is increasingly and recently used in assessing the quality and managing ecosystems is the biotope. This term may be regarded as describing the habitat and its component and dominant organisms. For example, middle estuarine salt marsh, mussel bed, or mudflat biotopes include the main physical features of the habitat and the dominant organisms (possible *Spartina, Nereis*, and *Mytilus* respectively). The presence or number of biotopes in an estuary, their size, and degree of functioning can then be used as a measure of the health of an estuary. As such, "biotope mapping" is increasingly used to determine these features and may be carried out at either a lower level (a Phase 1 survey) in which all assessments are done in the field and only the obvious organisms are noted. Alternatively, if greater data are required then a Phase 2 survey will be carried out whereby samples are taken for the full identification and enumeration.

7.6 Monitoring, surveillance, and survey design

In physiological terms the term "homeostasis" is used as the ability to cope with or absorb change by modifying the organism's functioning but without resulting in a long term, deleterious change. We may also think of estuarine populations, communities, and the ecosystem as exhibiting homeostasis—the ability to absorb and adjust to change without the effects being observed widely. As shown throughout this book, estuaries are highly variable systems and thus they may have the ability to absorb anthropogenic change as well as natural change and still function as expected. The aim of the estuarine scientist and manager is to detect whether those characteristics have been changed, thus a rigorous approach to monitoring is required. This is particularly important, as in some cases data produced by the monitoring may be used in legal discussions or proceedings.

In any discussion of monitoring, it is necessary to mention the quality of the data produced especially where those data are to be used widely. Because of this, there is an increasing need for analytical quality control (AQC) and quality assurance (QA). It is important to realize that a measurement is never exact but it can be precise and accurate within certain limits (types of equipment used, skill of the scientist, etc.). *Precision* is the *same* measurement or class achieved on each occasion that an observation is made, for example, the weight of an object on a particular balance or the same identification of a taxon by one taxonomist. In contrast, *Accuracy* is the *correct* measurement or class that is obtained on every occasion that an observation is made, for example, the weight of a standard calibration material or the correct identification of a taxon by any taxonomist. Hence, measurements can be precise without being accurate, for example, a badly maintained balance could be precise without being accurate. AQC/QA is designed to ensure that all the errors in measurement are known, accounted for, or minimized, for example, a laboratory has well-trained staff, well-documented methods, and good equipment. In this way, data obtained from one estuary can be combined with those from any other area thus allowing changes in one area to be put into a wider context.

As the widest example of an integrated study focused on a defined estuarine activity, Environmental Impact Assessments (EIA) are now widely required to be performed before permission is given for each estuarine activity. Those EIA should have a structured approach (see Glasson *et al.* 1999) and thus consider each component of the activity and each part of the natural system likely to be impacted. It should determine the impact or an activity at a place at a certain time and as performed in a certain way, that is, a site and time-specific set of effects. For example, the construction and operation of a pipeline into an estuary will involve clearing a channel, bringing materials to the site, perhaps dredging an area, building the pipeline, and discharging the waste from it. These aspects may affect different parts of the water column, estuarine bed topography and sediment, and faunal and floral populations. Hence the EIA should quantify the spatial and temporal impact of each part of the sequence on each natural component and then determine the size of that natural component before determining whether the activity will have a long- or short-term effect or will affect a small area or large area. Hence monitoring is required to quantify each of these aspects.

The monitoring is required to determine the spatial degree of effect, that is, the extent, and the temporal severity of effect, that is, the duration. The spatial degree can be the amount of area affected and the spatial importance of a component affected (Table 7.4). For example,

whether a component affected by an activity within an estuary is important in an international context, such as a population of overwintering wading birds, or only in a local context, such as an estuarine intertidal mussel bed. The size of the affected component needs to be determined, for example, the number of hectares covered by the mussel bed or whether the wading birds are >1% of the European population (i.e. internationally significant). The final conclusions from the matrix of impact (Table 7.5), after it has been applied to the activity, produce as an

Table 7.4 Terminology for indicating severity of an effect

Type of effect	Importance of ecological element					
	I	N	R	D	L	Nil
Negative						
Irreversible in long term	●●● ●●●	●●● ●●	●●● ●	●●●	●●	0
Reversible in long term	●●● ●●	●●● ●	●●●	●●	●	0
Reversible in short term	●●●●	●●●	●●	●	0	0
No known impact	—	—	—	—	—	—
Positive						
Short-term benefit	□□□ □	□□□	□□	□	0	0
Long-term benefit	□□□ □□	□□□ □	□□□	□□	□	0

Scales: I, International; N, National; R, Regional; D, District; L, Local and Nil, no known value.

Table 7.5 Example of use of impact scales for an estuarine pipeline construction

Source of impact			Impact "Target"			Nature of impact					Significance
Element	Activity	Aspect	Component	Potential size of target	Intrinsic ecological value	−ve irrev. long term	−ve rev. long term	−ve rev. short term	+ve short term	+ve long term	
Pipeline	Construction	Excavation	Mussel bed	0.5 ha	Nil	✓					0
			Fish	Small nos.	L			✓			0
			Bird community	1%, UK	N			✓			●●●
			Oystercatcher	<1%, UK	R			✓			●●

end point, a graphical and simple indication of degree of effect, whether positive or negative. This then needs to be communicated to interested bodies. For example, if it is decided that constructing a barrier will have a negative effect, which is irreversible in the long term on a population of international significance, such as overwintering wading birds, then a table of impact will indicate this by giving six filled circles.

Following this process, it is then necessary to indicate, in the EIA, whether any mitigation measures can be employed, for example, can a pipeline be constructed in such a way as to minimize an effect on the bed of the estuary. If mitigation is not possible, then managers will have to determine whether compensation measures can be produced, such as by creating or restoring a habitat elsewhere as compensation for a estuarine habitat, which had been destroyed by an activity.

7.6.1 Decision Tree

The nature of monitoring and surveillance required to answer the questions posed throughout this chapter may be defined using a Decision Tree. Surveillance is regarded as gathering data and then indicating change *a posteriori* whereas monitoring should relate to a pre-defined end-point, for example, the dissolved oxygen levels of waters required to protect estuarine fish migrations. Once a pre-defined limit of 5 mg l^{-1}, for example, is set then monitoring can be directed toward detecting the time and place where this is not reached.

Decision level 1: Definition of main questions and hypotheses
This first step will decide whether studies are carried out to provide answers to either general or precise questions, to test hypotheses, and to generate hypotheses. For example, many studies merely determine the usage of estuarine or marine intertidal and subtidal areas, the respective usage of estuarine and coastal areas and their associated freshwater areas, and the

partitioning by fish and bird consumers of available production. Studies are likely to be applied, for example, attempting to determine whether the indigenous fish populations in coastal areas are as they should be given the hydrophysical characteristics of the area and whether the benthic populations differ from similar areas. There may be the need to know the trophic structure, the estuarine foodweb, and its functioning. Finally, there is often the need to know how the populations have been affected by man's activities and whether Ecological and Environmental Quality Objectives (EQO) and Standards (EQS) are being met.

Decision level 2: Monitoring definition
In order to create a focused approach, the type of monitoring needs to be defined. For example, whether *Statutory Monitoring* is required and, if so, to determine who is responsible for carrying it out. This will follow identifying, for the human impact, the *Uses and users of the estuary or coast at risk and those who are interested in the results of the monitoring*. This determines where the demand is for the monitoring and why it should be carried out. Within estuarine and coastal management, this will produce a matrix of effects on the different users and uses.

The next decision is the *Detail of monitoring required*, which covers from the subjective and qualitative, through skilled-eye techniques to the fully quantitative and statistically rigorous surveys. A definition of the *Components at risk* will then indicate what aspects should be monitored, which in turn dictate the methods to be used. The scale of the monitoring then requires to be considered, first the *Spatial extent*, as the area to be monitored, which involves defining station positions, the strata to be sampled and often the station positions with regard to hydrographic characteristics. Second, the *Duration and Frequency* of sampling, which illustrates the temporal component and the length of campaign. Finally, the sampling may be required to detect changes in the biological and environmental components, hence the *Degree of change expected/tolerated* should be defined, thus dictating the number of samples and the degree of replication.

Decision level 3: Types of survey required/desired
There are many types of survey:

• background surveillance, to improve information on natural features and their inherent spatial and temporal variability
• condition monitoring, to determine if the status of the estuarine individuals, populations, and communities are in a pre-defined state, as expected by the ambient conditions
• compliance monitoring, whereby standards and targets have been set, such as the amount of contaminant in fish flesh, and require to be met.

In many cases, these surveys relate to the determination of impacts as the result of human activities such that they require:

• an exploratory survey, such as a widespread scanning of an area, the use of published and unpublished information and the accompanying desk-study
• a baseline study, which may have a large spatial coverage, perhaps with low intensity methods and few, if any, replicates, in order to define the area for further, detailed study or where actual or potential problems may occur
• an ongoing survey, which where possible should be a time-defined study, statistically robust and with a frequency of sampling relevant to the timescale of the processes under study

In the case of the detection of the effects of human activities, the above surveys are likely to include the BACI–PS (Before-After-Control-Impact Paired-Series) approach. The replication of sampling sites and occasions will have to be sufficient to detect the type of change likely.

Decision level 4: Associated parameters/ integrated monitoring
In order to fully interpret the biological information, associated information is required

such as:

• physical monitoring, involving characterizing the hydrographic and sedimentary regime
• chemical monitoring, for example, temperature, salinity, prevailing water, and sediment contamination
• community biology, such as related components, either as a single component (hard benthos, soft benthos, hyperbenthos, nekton, plankton, etc.), or a combination (two or more components)
• ecotoxicological/individual health, such as the presence of contaminant detoxifying mechanisms, pathological anomalies, bioaccumulated contaminants
• socioeconomic aspects, such as the degree of commercial fishing, type of bycatch.

Decision level 5: Methods to be used in monitoring
It is emphasized that once the aim of the monitoring has been clarified then the methods for that monitoring should follow easily. All methods have advantages and disadvantages in assessing and quantifying the actual and potential degree of change in the component(s) of interest. As field and laboratory studies are expensive, the intensity and design of study should reflect the size of the actual or potential changes. There are many designs of survey and each is suited for answering specific questions and for use in various estuarine habitats.

7.7 Final comments

This chapter has indicated that there is a suite of methods for determining change in estuaries, both due to natural perturbations but mainly due to human activities. Each of those methods will detect a type of change and may be valuable for a particular habitat, biological component, even individual species, or type of pollutant or other stressor. The different techniques have advantages and disadvantages, costs and benefits and usually a combination of techniques will be required—however, it should always be remembered that monitoring is expensive. We will never have a single nor fool-proof method for detecting change and often as we carry out more assessments then we merely describe the large-scale inherent spatial and temporal variation in estuaries.

As a summary it can be suggested that with any method of stress assessment, a four-stage process is required:

Explanation of the rationale behind change,
Evaluation (quantitative) of the nature of change,
Interpretation of the effects shown,
Prediction of the effects under a given stress.

The last of these is particularly important— many estuarine scientists will be asked, *de facto*, to predict an effect—what will happen when a bridge is built, where will the effluent go and what will be affected, what will happen if an estuarine channel is dredged, and so on. We will have to give an answer while at the same time remembering the limitations of our knowledge and our tools for studying estuaries. We have to remember that all methods have problems, many are in early stages of

validation and are not necessarily suitable as working techniques. In many cases, the sources of variability in method has not been determined or quantified. Underwood and Peterson (1988) take the view that:

Utilizing measures of pollution at different levels of biological organization represents a sensible strategy for environmental scientists because these measures serve different purposes. The organismic and sub-organismic measures potentially provide the earliest warning of possible future deterioration and may also be the most sensitive measures of pollution. In contrast, measures on communities may provide a better indication of the consequences of that pollution to processes of economic and societal value vested in the marine ecosystem. A complete assessment of any episode of pollution must include accurate measures of the biological effects of that pollution at a number of different scales. No one measure can satisfy all the requirements of those who must make decisions about potential environmental, economic and social impacts of pollution.

The management of estuaries, the legislative, socioeconomic and administrative procedures, protocols and considerations will be discussed in Chapter 8.

CHAPTER 8

The management of estuaries

8.1 Introduction

The main objective of estuarine management is to devise a framework within which man may coexist with nature. Here we have divided estuarine management into three broad areas: policies and philosophies, planning and designations, and practice (after Carter 1988), although these are all interlinked. Policy reflects the underlying approaches and frameworks in the way we manage estuaries, which is related to the political and administrative framework through which estuarine management is regulated, be it by legislation or education. Planning is used for the process of resource allocation, be it by ecological or economic yardsticks. We show how the management of certain components (species, habitats, whole areas, industries, catchments, etc.) are interlinked. Practice covers the means by which we can derive and implement decisions, such as the organizations and stakeholders involved in estuarine management, the legal framework under which they operate, and the outcome of their actions. The latter may include mitigation and compensation mechanisms, or the introduction of technologies such as building new sewage-works or treating industrial effluent. All of these aspects are interrelated and each follows from each other.

The management of estuaries has to cope with three fundamental paradoxes. First, the majority of the world's major cities are located alongside estuaries, yet, for most of the inhabitants of those cities, estuaries are the most natural wildlife habitat that they encounter. Second, most of the major estuaries of the world are to some degree modified by Man or polluted, yet, in many countries there are more estuarine nature protected areas than for any other habitat. Third, many estuaries receive high levels of organic wastes from sewage, farming, and other industries, yet estuaries are amongst the most productive natural ecosystems known. Estuaries are, however, moderately diverse and are highly variable and this may allow them to absorb the low-level effects of human activities without major changes.

8.2 Policies and philosophies

Given the natural features of estuaries and the demands put on them by Man's activities, management has to reconcile these features but to allow them to coexist. As indicated in Chapter 6, the multiple stressors in estuaries each have to be managed and in most countries that management has always been on what we may call a *sectoral* basis, that is, the management of each activity (fisheries, recreation, navigation, etc.) in turn. Since the late 1980s Man has realized that in order to have a healthy and successful ecosystem in which human uses and users may be allowed, accepted, or tolerated then there is the need to move from the sectoral to the *holistic* basis. In this it is accepted that all the components of the estuarine ecosystem are interlinked and so all of these, the physico-chemical and biological, have to be managed in such a way that the system is sustainable. All of this may be called "the ecosystem approach."

Box 8.1 Estuarine environmental management

What are the "legitimate" uses of the estuarine system?
  (What are the accepted uses?)
What are the human demands on the system?
What/where are the conflicts?
  (Do activities conflict spatially/temporally?)
Can the physical/biological system cope?
  (Can adverse effects be detected?)
Should/can the activity be stopped?
  (Can the conflicts be tolerated?)
Would zoning/designation resolve potential/ actual conflicts?
Are mitigation measures feasible?
  (Should mitigation measures be carried out?)

The basis of this is to create a framework for management to decide what we want from estuaries and whether we can allow activities to occur (see Box 8.1). We use the term "legitimate" activities because in general, especially in developed countries, these activities have been sanctioned and licensed.

This chapter will also show that recent changes in our concepts and practices in estuarine management have shown a shift in their underlying basis. The main features in estuarine management, summarized in Box 8.1, show the importance of types of information, the need for determining priorities, and the tools that are available (laws and protocols). Management initiatives previously developed for use on land such as the planning process are now being applied to the aquatic environment. Initiatives developed in freshwaters, such as habitat and species management, have been applied to inshore/estuarine areas and more recently to offshore areas. With a progress in management from the land, to freshwaters, to estuaries and now to coastal and offshore areas, we have to cope with the problem of some aspects but not others being quantified, and the limitations of scientific knowledge. The management of a piece of grassland can be relatively straightforward as it may link very little with nearby habitats, let alone those many kilometres away. In contrast, in estuaries we have to be aware of developments not only within the estuary but also in the catchment and at sea, and within an environment that is changing on a tidal, diurnal, weekly, lunar, seasonal, and annual basis.

In addition to these changes, we increasingly have to consider Man's aspirations—the increasing environmental awareness of the population. The latter may be for aesthetic concerns, as in developed countries, such as the value of having high numbers of so-called "charismatic megafauna" such as seals, dolphins, or wading birds, or because of the realization that a healthy environment is necessary for good food production and can be economically necessary. It is possible to put an economic and social value on the wealth creation aspects of estuaries—providing fishes, land for building, or as navigation waterways. A fundamental difficulty, however, in assessing the importance of estuaries is in how Man can put a value on particular species and habitats and how, with a conservation emphasis, we can put a value on aesthetic aspects.

Because of the complex nature of estuaries, their users and uses and the effects of these, there is a need for what we may call a multi-sectoral/multi-organizational/multi-effects/multiuser/multiuse approach. In environmental management, we often have a limited ability to cope with such a complex situation and so, as will be shown here, we usually need to break down the problem into its parts.

The most difficult part of managing an environment is to decide what is required of that system, in this case an estuary, by its users. For example, do we want an area to support wildlife, provide fish and shellfish, allow navigation, receive waste or support land for housing or industry, or do all these things? Before it is managed, it is necessary to determine the estuary's condition, the change in that condition from what is expected if an activity goes ahead and to decide what is the desired condition for the estuary. Because of this, there is a need for indicators to describe the status of the ecological/ecosystem elements, as in the "DPSIR" approach (Chapter 6), which summarizes the source of problems, the causes and effects against the background characteristics, and the responses required by Man in order to manage the estuarine area. The options for sustainable and cost-effective management then have to fulfil six tenets: being environmentally sustainable, economically viable, technologically feasible, legally permissible, socially desirable, and administratively achievable. For example, dredging of an estuarine waterway will be required by society in order to make the area suitable for navigation but it has to be carried out using appropriate and cost-effective methods, in accordance with local laws and international obligations but without having an adverse and unacceptable environmental effect. This chapter aims to illustrate many of these features.

Over the past half century, there has been an evolution of the thinking regarding environmental protection relating to pollution and discharges into estuaries. In the 1960s there was recognition of the problem, following public awareness as indicated by Rachel Carson's book *Silent Spring*. The 1970s saw an increase in ecological monitoring and the creation of end-of-pipe controls with the use of long-sea outfalls and preliminary treatment. The 1980s had some better treatment and tighter end-of-pipe controls whereas the 1990s led to the "holistic approach." The latter brought in Integrated Pollution Control (IPC) and Environmental Management Systems (EMS), and clean technologies for industries. It acknowledged that pollution moves across land, air, and water and the need to stop pollution at source rather than merely treat it after it had been produced. The 2000s completed the cycle by showing a more complete recognition of the problem—the acknowledgement that an ecosystem approach was required in which all features were considered and managers and

scientists had to be aware of a large number of aspects. It was recognized that although individual activities should be controlled, the main aim is to manage the whole environment, and to protect habitats and species and the functioning of estuaries. This created an overall philosophy that incorporated sustainability, precautionary action, the integration of all environmental aspects, and democratization (wide consultation, open, fully accountable) (Table 8.1).

In managing estuaries, we can take what may be called "bottom-up" or "top-down" approaches. For estuaries, this is from the (bottom-up) management of species and habitats, through activities, and whole estuaries to (top-down) management of catchments and coastal sea areas. For legislation and administration, this is from (bottom-up) local and regional initiatives, national laws and bodies, supranational ones such as the European Union (EU), to (top-down) conventions and agreements covering regional seas and global aspects.

8.3 Planning and designations

8.3.1 Management and protection of species and habitats

There is an increasing awareness of biodiversity as the protection of all members of the biological community although often these are the high profile species, the rare, fragile, and endangered species or the species important in the overall functioning of the system. Management initiatives are then required to protect each of these species.

The International Union for Nature Conservation (IUCN) has created lists (published as the Red Data Books) of endangered species, which may then form part of the species conservation measures created by countries. In the case of European estuaries, these aspects have been incorporated into the European Commission's Habitats Directive, which in Annex I give habitats for designation and

Table 8.1 Underlying themes in estuarine management

Main feature in environmental management	Relevance to estuaries
Sources of legislation—national, supranational, etc.	To include specific laws that relate to estuaries or have been implemented to include estuaries; to determine whether estuaries are "controlled waters" and identify the competent authorities
Types of agreement, protocols— national, supranational, global	To include agreements, from inside and outside states, which may or may not lead to estuarine legislation
Identification of priorities for action	To determine which are the greatest potential or actual problems within an estuary, such as which pollutants are important, what habitats and species should be protected, what water characteristics need to be assessed
Authorization for potentially harmful action—licensing, etc.	To decide which of those priorities for action require a licence, for example, do estuarine industries require integrated pollution control authorizations, or which dredging schemes require permits?
Precautionary principle, action	To decide which of the activities should not be permitted on the basis that they are likely to cause harm to the estuary
Impact assessment and action	For each activity which has the potential to harm the estuary, to be subject to an EIA, that assessment must include all anthropogenic and natural characteristics outside of the estuary
Roles of physics, biology, and chemistry in assessment and control and the integration of environmental aspects	The need to ensure that all of the interlinked estuarine processes are included and to use information from one discipline (e.g. hydrography) to inform decisions about another (e.g. biology)
Role of monitoring and surveillance	To ensure that the surveillance of the estuarine features is carried out (i.e. determining spatial and temporal trends), but also that monitoring *per se*, in which the assessment is carried out against a preconceived threshold/level of acceptable change, is performed in a cost-effective manner
Sustainability for use vs non-use	To realize that within a working estuary that supports many economic activities such as navigation and industry, it is necessary to ensure that they are carried out in a sustainable manner (i.e. one activity does not jeopardize another either now or in the future). Similarly, non-(economic) use such as maintenance of wildlife also needs to be sustainable
Carrying capacity of natural and anthropogenic features	It is argued that an estuary has a carrying capacity for each aspect, for example, the number of wading birds that it supports, or the amount of waste discharged into it without creating harm, and that management has to be geared to ensuring that these occur
Democratization, accountability, openness	If there are changes to an estuary, management of its features, or the awarding of permits for activities, then all stakeholders have to be involved
Polluter pays principle— monitoring, clean up, prevention	Estuarine users, such as dischargers putting waste into an estuary or navigation authorities carrying out dredging, are responsible for the effects to be monitored, clean up to be effected, new technology to be used
Classification schemes, performance indicators	The estuarine areas have to be classified according to their nature and their quality, and the amount of estuarine area within each quality class can be used as a performance indicator to determine whether management is being effective
Environmental management systems, clean technologies, waste minimization.	Within a cost–beneficial and environmentally sustainable framework, industries, and developers are made to be responsible for ensuring that environmental problems either do not occur, are mitigated (made less severe) or compensated (that the loss of one habitat is compensated for by creation elsewhere)

protection. Its Annex II lists species important in a European context and which may be requiring protection because they or their habitat are endangered. While there are few estuarine (and nearshore marine) species on the list, there are 12 species of fishes (Table 8.2). Most of these are diadromous species, that is, they migrate between freshwaters and the sea, and so any barriers, whether physical in the case of weirs and barrages, or chemical, such as dissolved oxygen sags, will endanger the population. Because of this, European Member States have to produce management plans to protect these species. Hence there is the increasing view that in estuaries and other marine environments, in order to protect the species it is preferable to look after the habitat and ensure that a species is not taken excessively, as in fishing. Following this, the species will look after itself. Once the species requiring protection have been designated then species-specific Biodiversity Action Plans (sp-BAP) can be created detailing the measures to be taken to protect the species. Wider measures may be required for species subject to commercial fishing; for example, No-Take Zones may be designated in certain areas.

Estuaries are not highly diverse areas in terms of numbers of species but they have a unique and important functioning and they have habitats not found elsewhere—such as mudflats, mangroves, salt marshes. There is thus the need to protect the estuary as a habitat or the habitats within estuaries and to protect habitats supporting large numbers of a few species (or even a single species). Because of this, habitat-specific Biodiversity Action Plans (hab-BAP) are created to indicate how much of a habitat is desirable and how much of that habitat should be protected on a national scale. Hence, the European Habitats Directive Special Areas of Conservation (SAC) have designated a type of habitat (e.g. sand banks) or a group of habitats (estuary, coastal embayment) thus providing protection for those habitats. Of particular relevance to

Table 8.2 The species of European estuarine and migratory fishes deemed to be fragile, threatened, or endangered

Species	Common name/type	European range	Comments
Lampetra fluviatilis	Lampern	Baltic to N. Mediterranean	Spawning in spring in UK, summer further north
Petromyzon marinus	Sea Lamprey	N. Africa to Arctic waters	
Acipenser sturio	Sturgeon	N. Africa to Arctic, more prevalent in north	Spring spawning in France
Acipenser naccarii	Adriatic Sturgeon	Only Adriatic	Very rare and verging on extinction
Aphanius fasciatus	Cyprinodontid	Mediterranean E. of Mallorca	Restricted, warmer waters
Aphanius iberus	Cyprinodontid	S. and E. Spain, W. Algeria	
Alosa alosa	Allis Shad	Canaries to southern Norway, W. Baltic and W. Mediterranean	More common Ireland southwards; spawns April–May
Alosa fallax	Twaite Shad	Canaries to Iceland, Baltic and Mediterranean	Rare in N. Sea and Baltic; May–June spawning (BI)
Salmo salar	Salmon	N. Portugal to Arctic	Winter spawner, northern species
Coregonus oxyrhynchus	Houting	W. Baltic, S. North Sea	Rare in southern parts
Knipowitschia panizzae	Lagoon Goby	Adriatic	Very restricted area
Pomatoschistus canestrini	Gobiidae	Adriatic	Very restricted area

estuaries is the European designated Special Protected Areas (SPA) for the protection of the habitat for supporting birds. Similarly, sites designated worldwide under the Ramsar Convention are aimed at protecting wetlands as bird habitats.

Although the protection of species and habitats is the main aim in management, it is likely that human activities will affect those and so mitigation measures have to be employed to minimize effects. Where this is not possible, for example, if a new dock removes mudflats, then compensation will be required such as the creation or restoration of habitat elsewhere in order to replace any habitat lost. This feature has now led to the process of "habitat-banking" in which a developer may buy land elsewhere in the estuary prior to development at one site taking place. The management of habitats on a larger scale in the United Kingdom is now being proposed through the creation of Coastal Habitat Management Plans (CHaMPs), which may indicate the overall management of habitats in relation to those affected by the estuarine activities.

8.3.2 Management and licensing of activities

In most areas, permission from a relevant statutory authority is required in order to perform an activity. That permission may be in the form of a legally binding document, which may be called a permit, a licence, an authorization, or consent. For example, in the United States, a NPDES (National Pollutant Discharge Elimination Scheme) *permit* is required before discharge of a land based effluent, whereas in the United Kingdom an *authorization* is required for a complex industrial plant to allow the integrated management of discharges to the air, into the estuarine waters, or onto land. A *consent* to discharge is required by a sewage treatment plant in order to discharge its effluent into nearby water course, whereas a *licence* is required to dump dredged material taken from a navigation channel.

Any noncompliance with the permit or licence will lead to legal proceedings. The main features given within a licence are indicated in Boxes 8.2 and 8.3.

Different approaches can be adopted to bring about pollution control. Either, an Environmental Quality Objective (EQO) policy, usually linked to an Environmental Quality Standard (EQS) policy, or a Uniform Emission Standard (UES) policy. The essence of the EQO approach is that an estuary (or any water body) can disperse, degrade, and assimilate pollutants that enter it. It is based on the concept of "assimilation capacity," which is the capacity of a body of water to cope with the effluents discharged into it. Apart from any chemicals, which bioaccumulate, it assumes that "the solution to pollution is dilution!" The EQO approach is concerned to define the overall conditions of the habitat and then to select for each estuary a set of permissible operations, or levels of pollutants, discharged in order to achieve that condition. The EQSs are

---

Box 8.2 Conditions imposed within a Consent (Licence/authorization/permit) for a discharge into estuarine waters

(1) types of outlet and number of diffuser ports;
(2) source of material discharged (type of industry);
(3) pH, BOD_5, suspended solids, temperature;
(4) emission levels (as concentrations—amounts per liter) (based on toxicity information + ability of receiving areas to degrade/disperse/assimilate);
(5) maximum loadings of key parameters (amount per day, month);
(6) volume of discharge per day and per hour;
(7) position of sampling points within plant;
(8) information on record keeping and good practice;
(9) toxicity-based element (Direct Toxicity Assessment).

Box 8.3 US System—National Pollutant Discharge Elimination Scheme (NPDES) (Clean Water Act—Water Pollution Control Act, issued by EPA)

Permits given stipulate:

(1) effluent characteristics and type of treatment to effluent;
(2) type of discharge system (diffuser type);
(3) dimensions of mixing zone and characteristics and concentrations at edge of mixing zone;
(4) type of testing of physical dispersion of effluent;
(5) oceanographic/hydrographic monitoring;
(6) other physical parameter monitoring (temp., weather, salinity);
(7) benthic communities in receiving areas;
(8) fate and effects of pollutants, especially persistent contaminants.

State and federal licences needed, former can be stricter than latter.

---

attempt to manage all activities in tandem, maintain all habitats and species, protect the natural functioning, and involve the stakeholders. A starting point, however, should be the underlying aspirations or objectives for the estuary.

These may be summarized as EQOs, such as those adopted for estuaries in the United Kingdom, in that the quality of the estuary should allow:

1. The protection of all existing defined uses of the estuary system.
2. The ability to support on the bottom the biota necessary for sustaining sea fisheries.
3. The ability to allow the passage of migratory fish at all stages of the tide.
4. The estuary's benthos and resident fish community and populations are consistent with the hydrophysical conditions.
5. The levels of persistent toxic and tainting substances and microbial contamination in the biota should be insignificant and should not affect it being taken by predators, including man.

The use of the information can be incorporated into Environmental or Ecological Quality Objectives (EQO or EcoQOs), as statements, which can then be tested to determine whether an area has natural or man-affected characteristics (Elliott 1996). Those objectives can then be framed as null-hypotheses that act as basis for the sampling and subsequent analysis. The above list has been chosen mainly on biological grounds, but there could be an EQO for each defined usage of the estuary. For example, water contact sports might require a different list.

The uses which are recognized in the first EQO are: disposal of effluent, commercial fishing and angling, nature conservation, bathing, boating and water skiing, tourism and amenity, navigation, and water abstraction for industrial or agricultural purposes. All of these uses, except the disposal of effluent, require the estuary to be as free of pollution as possible. In order to meet that objective the

the levels of permitted discharges that are selected for a particular estuary. Thus the EQOs are the stated aims for an estuary, whereas the EQSs are the conditions used to bring about those EQOs. The alternative UES approach is to establish a national or international standard for the discharge of a pollutant, and then to apply those standards to every discharge from a specific industry or process, irrespective of the location of the discharge. The emphasis in a UES policy is clearly on the uniformity of discharge from different sources or locations.

The holistic management of estuaries and the integration of all uses and users has increasingly led, since the early 1990s, toward the creation and adoption of Estuary Management Plans and there are many notable examples of these in the United States, Europe, and Australia with the latter country combining these for its estuaries to give a National Estuaries Management Programme. These disposal of effluent must be controlled, either by the reduction in volume, or by an improvement in quality of the waste so that it does not impede the other uses. At the mouth of the estuary, where there is plenty of water, this objective might be attained by better dispersion of existing waste through the construction of a long sea outfall waste pipe. In the inner estuary where there is less dilution water available, the attainment of the same objective might involve the construction of new sewage-works or treatment plants. The standard of effluent permitted may thus be different in different parts of the estuary. In order to achieve the defined EQOs, a higher EQS would generally have to be applied to a sewage or industrial discharge at the head of an estuary, than to one at its mouth.

The second objective requires that the estuary be of sufficient quality to sustain the diverse components of the estuarine ecosystem that we have described in earlier chapters. In particular, no discharge to the estuary should remove the fauna in the vicinity of the discharge, and reclamation and other engineering schemes that destroy the benthic habitat should be avoided.

The third objective, to allow the passage of migratory fish, such as salmon or eels, at all states of the tide, is designed to ensure that the oxygen concentration, as a measure of water quality, is greater than 50% saturation at both high and low tide, and in both summer and winter. In many estuaries, this objective is readily attainable in winter at high tides, when there is a large volume of water, but more difficult to attain in summer at low tide, when there is a smaller volume of water and there will be less oxygen present due to the water being warmer or more saline (see, for example, Fig. 6.4). This objective must therefore be defined in terms of the most problematic season or condition, not the most lenient.

The fourth objective is to ensure that the elements of the biota are as expected given the area's hydrophysical characteristics, its water movements, and substratum type. Thus if a low energy area is protected and maintained then it will support a mudflat containing large numbers of prey for overwintering birds and resident fishes.

The last objective, of those given here, is designed to eliminate contamination of, for example, shellfish by microbes, such as fecal coliform bacteria, which may be injurious to any person eating the shellfish. It also aims to eliminate contamination of fauna by any persistent accumulative pollutant, such as heavy metals (especially mercury or cadmium) or Organohalogens that could taint them, affect predation, or harm anyone eating them.

In certain cases, the meeting of these objectives is determined by testing for specific and numerical standards, as EQSs, which may be created for licences or protocols. The EQS values may be defined as the maximum allowable concentration, and/or the 95%ile value (95%ile = the value that is not exceeded for 95% of the values taken), and/or the annual average. These values may be included within the licence or the discharge conditions may be set allowing for the dilution, dispersion, and assimilation within the receiving waters to ensure that the EQS are not exceeded.

Toxicity testing (see Chapter 7) has a role for estuarine management in helping to determine the permitted levels of toxicant in the discharge. For example, 1/10 or 1/100 of the 96 h LC_{50} may be used as the "safe level" for the discharge of a toxic effluent, these factors being termed "application factors." Such application factors have been criticized, because exposure to sublethal concentrations of a pollutant may influence physiological processes such as reproduction or development, but not cause death. Despite such valid criticism, the acute toxicity test and application factors remain in use, since such tests are easy and reliable to perform, and provided that the most sensitive life stage of an animal is used, do give valuable guidelines for determining priorities in the management of industrial effluents discharged to estuarine ecosystems. Toxicity testing has long been used in determining the levels of single compounds for discharges. All discharges, however, convey a mixture of materials, so it is necessary to licence the complex effluent and, because of synergistic and antagonistic effects, there is an increasing use of a direct toxicity elements in the licence (called toxicity based consents or Direct Toxicity Assessments). The toxicity element would therefore be a clause in the licence whereby the whole effluent would be tested (Whole Effluent Testing) and deemed to fail if its LC_{50} was likely to be breached after dilution in the receiving waters.

When the policies of management in estuaries are implemented as planning objectives, the concepts of EQO, EQS, and UES are perhaps not as different as they first appear. In practice they usually work together, in that the estuarine manager needs to specify the permitted standards in any effluent entering the estuary, in order to achieve the objectives, which have been set for that estuary. Whether the standards set are uniform in all countries, or parts of that country, is essentially a political decision. The role of the estuarine scientist is to advise the politicians as to what standards are required, either in the effluent, or the receiving waters, so that the estuary is fit for its uses. In the context of estuarine water quality, the concentration of a particular constituent in a discharge may not be the most important factor; it is the rate of entry to the estuary that matters. This is obtained by multiplying the concentration by the rate of flow to give the "load." The assessment of loads is, however, often difficult since many of the sources of pollution are subject to wide variations in both flow and composition (Gameson 1982). As the effluent load enters the estuary it is usual to define a "zone of acceptable ecological impact," or a "mixing zone," which is an area where dilution and dispersion occurs. The zone should be as small as practicable, and some latitude in the EQS for this area may often be deemed acceptable.

Pollution discharged into an estuary is managed by determining the state of the estuary, the quality of its waters, and the condition of its inhabitants at all times of the year. Discharges entering the estuary from industries or domestic sewage-works or the diffuse sources from the river input should be identified and analyzed and, in the case of the anthropogenic discharges, licensed by the relevant authority. The characteristics of the licence (Boxes 8.2 and 8.3) control the levels entering the estuary and ensure compliance with international and national standards and agreements, such as whether the discharge contains any of the priority substances of concern as well as the local situation of the discharge. The licences can then be reviewed at intervals as technology improves and knowledge about the fate and effects of pollutants becomes available.

The US state and federal policy has generally followed the EQO/EQS approach. In the 1970s and 1980s, the UK government also followed an EQO/EQS approach, but with the adoption of the North Sea declaration following the North Sea Ministerial Conference in 1987, and the adoption of the precautionary principle, it has moved to a UES policy for those substances regarded as causing the greatest problems, that is, being the most polluting. These substances, given on the so-called "Red List," will be controlled by cutbacks in all emissions irrespective of the body of water to which they are discharged. The precautionary principle could lead to zero emissions given that all substances are likely to have an impact sooner or later, at some place and in some concentrations. The precautionary principle may be defined as a philosophy whereby "preventative action should not be postponed even if the causal link between an activity and the impact has not yet been fully confirmed."

Some EQSs for estuaries and seas in Europe are listed in Table 8.3. Such lists are ultimately based on the evidence of the toxicity of the different substances with mercury clearly

Table 8.3 EQS for Europe. All standards expressed as microgram per liter (=parts per billion), in receiving waters (dissolved), except where indicated

Parameter	Estuary	Coastal sea
Dissolved oxygen	>55%	
pH	6.0–8.5	
Ammonia	<0.025 mg l^{-1}	
Cadmium	<5.0	<2.5
Copper	<5.0	<5.0
Mercury	<0.5	<0.3
Arsenic	<25	<25
Nickel	<30	<30
Lead	<25	<25
Zinc	<40	<40
Chromium	<15	<15
Iron	<1.0 mg l^{-1}	<1.0 mg l^{-1}

Source: After FRPB (personal communication) and Sayers (1986).

regarded as the most toxic heavy metal. Such standards for EQS of waterborne contaminants may not necessarily reflect the potential harm to the fauna in the sediments. For example, mercury goes out of solution very quickly as it is adsorbed onto suspended particles, which then accumulate in sediments. There is therefore a move toward future EQSs being based on sediment concentrations. It should also be remembered, as discussed in earlier chapters, that the toxicity of any single compound may vary greatly according to environmental factors within the estuary such as temperature and salinity, or may vary when the single compound is part of a cocktail of many chemicals.

A common theme in international policies to regulate the discharge of toxic substances to marine and estuarine waters is the recognition of two types of contaminants in order to prioritize control. First a "Black" List or List I of substances, which should be banned or substantially eliminated from discharges, and second a "Grey" List or List II of substances the discharge of which may be permitted in

Table 8.4 The Black List—substances which should be banned or substantially eliminated from discharges. The list is of the "Paris" Convention and "List I" of the European Dangerous Substances Directive

Priority substances for pollution control

List I—certain individual substances (see below) selected according to their toxicity, persistence, and bioaccumulation potential, with the exception of those which are biologically harmless or which are rapidly converted into substances which are biologically harmless

1. Organohalogen compounds and substances which may form such compounds in the aquatic environment
2. Organophosphorus compounds
3. Organotin compounds
4. Substances in respect of which it has been proved that they possess carcinogenic properties in or via the aquatic environment[a]
5. Mercury and its compounds
6. Cadmium and its compounds
7. Persistent mineral oils and hydrocarbons of petroleum origin
8. Persistent synthetic substances which my float, remain in suspension or sink and which may interfere with any use of the waters

[a] Where certain substances in List II are carcinogens, they are included in category 4 of this list.

carefully controlled quantities. These priority lists (Tables 8.4 and 8.5) are based on Annexes of the 1992 Paris Convention (formerly the linked Oslo and Paris Conventions, OSPAR) which has been adopted by many countries, to control land-based discharges and dumping into the NE Atlantic area. This approach is similar to the global London Convention (formerly the London Dumping Convention) of which the United States is a signatory and thus has adopted the substance of these lists. The Black List is also List I of the European Dangerous Substances Directive, and the Grey List is List II of that directive. These directives are the "parent" directives setting out the framework for the control of waste discharges in Europe, and are updated by the addition of "daughter" directives as new compounds are identified as

Table 8.5 The Grey List. Substances whose discharge may be permitted in carefully controlled quantities. The list is of the "Paris" Convention, and "List II" of the European Dangerous Substances Directive

List II contains: substances belonging to the families and groups of substances in List I for which the limit values (referred to in Article 6 of the Directive) have not been determined; certain individual substances and categories of substances belonging to the families and groups of substances listed below; and those which have a deleterious effect on the aquatic environment, which can, however, be confined to a given area and which depend on the characteristics and location of the water into which they are discharged.

Families and groups of substances referred to in the second indent

1. The following metalloids and metals and their compounds:

1. zinc	6. selenium	11. tin	16. vanadium
2. copper	7. arsenic	12. barium	17. cobalt
3. nickel	8. antimony	13. beryllium	18. thalium
4. chromium	9. molybdenum	14. boron	19. tellurium
5. lead	10. titanium	15. uranium	20. silver

2. Biocides and their derivatives not appearing in List I
3. Substances, which have a deleterious effect on the taste and/or smell of the products for human consumption derived from the aquatic environment, and compounds liable to give rise to such substances in water
4. Toxic or persistent organic compounds of silicon, and substances that may give rise to such compounds in water, excluding those that are biologically harmless or are rapidly converted in water into harmless substances
5. Inorganic compounds of phosphorus and elemental phosphorus
6. Non-persistent mineral oils and hydrocarbons of petroleum origin
7. Cyanides, fluorides
8. Substances which have an adverse effect on the oxygen balance, particularly: ammonia, nitrates

being harmful (e.g. pentachlorophenol, PCP). The United Kingdom, together with other countries whose estuaries enter the North Sea, in 1987 adopted a "Red" List (Table 8.6) of substances whose control is a priority, which will be controlled by cutbacks in all emissions irrespective of the water body to which they are discharged.

To control the discharge of material, which may harm shellfish, EQSs for waters with shellfish populations have been designated in both United States and Europe (Tables 8.7 and 8.8). Unlike the standards previously listed, these standards include bacterial or microbial contamination. Standards, again based on bacterial or microbial contamination have also been adopted for beaches used for public bathing (Table 8.9).

The adoption of EQOs has been criticized by several agencies and governments as being

Table 8.6 The Red List. Substances whose control is a priority, which will be controlled in the North Sea, and its estuaries, by cutbacks in all emissions irrespective of the water body to which they are discharged

Mercury	Triorganotin compounds
Cadmium	Dichlorvos
Gamma—Hexachlorocyclohexane	Trifluralin
(Lindane)	Chloroform
DDT	Carbon tetrachloride
Pentachlorophenol (PCP)	1, 2-Dichloroethane
Hexachlorobenzene (HCB)	Trichlorobenzene
Hexachlorobutadiene (HCBD)	Azinphos-methyl
Aldrin	Fenitrothion
Dieldrin	Malathion
Endrin	Endosulfan
Chloroprene	Atrazine
3-Chlorotoluene	Simazine
PCB (Polychlorinated biphenyls)	

Source: North Sea Ministerial Conference 1987.

Table 8.7 Shellfish-rearing waters standards for the State of Virginia, USA

Parameter	
Dissolved oxygen	4–5 mg l^{-1}
pH	6–8.5
Total coliforms	70–100 ml^{-1}
Faecal coliforms	14–100 ml^{-1}
Mercury	0.1 µg l^{-1}
Cadmium	5.0 µg l^{-1}
DDT	1.0 ng l^{-1}
Dieldrin	3.0 ng l^{-1}

Source: Environmental Protection Agency (EPA) of the United States.

Table 8.8 Environmental quality standards for Europe on the quality of water required by designated shellfish waters

Parameter	Mandatory EQS	Notes
PH	7–9	
Coloration	Deviation <10 mg Pt^{-1}	
Suspended solids	Increase <30% background	
Salinity	<40% increase, <10% background	
Dissolved oxygen	<60%, average >70% saturation	
Petroleum	No visible film	
Hydrocarbons	No harmful effects	
Organohalogens	No harmful effects[a]	
Trace metals	No harmful effects[a]	
Fecal coliforms	<300/100 ml	Applied where live shellfish directly edible by man
Substances affecting taste	Below taste threshold in fish	Applied where presence is presumed

Element	Annual mean conc.	Substance µg l^{-1} wet wt	Flesh conc. ng g^{-1} where lipid conc. ≤1%
Ag	0.3	Dieldrin	50
As	20	DDD[b]	100
Cd	1.0	DDE[b]	100
Cr	10	DDT[b]	100
Cu	5.0	HCB	100
Hg	0.1	aHCH	30
Ni	5.0	yHCH	30
Pb	5.0	PCB	1000
Zn	10.0	Toxaphene	1000

[a] Trace metal and organohalogenated substances: levels below which harmful effects will not occur.
[b] Sum of DDT 250 ng g^{-1}.

Source: EC Directive (79/923), supplemented by guidelines adopted for Scotland. Mandatory means that the values given should not be exceeded.

unjust. The opponents of EQOs say that they unfairly penalize a town or industry situated inland or near the head of an estuary, where there is little scope for dilution of an effluent, against a town or industry situated on a sea coast, or near the mouth of an estuary, where abundant scope for dilution is available. Thus more lenient EQSs might be applied to the discharge from an industry at the estuary mouth. The UES approach has been adopted to overcome this criticism as a set of numerical standards to which all industries or towns must conform irrespective of their location. UESs are particularly suitable for situations on a continent where a river may pass through several countries or states, such as the Rhine or the Danube, so that an industry in one country is not unfairly treated compared to similar industries in other countries.

As a recent addition to the adoption of the EQO/EQS philosophy, Australia and New Zealand through their joint Water Quality Guidelines have created a framework (Box 8.4). This creates different water quality standards and reference sites for different ecosystem types, including estuaries, but also for different ecosystem conditions. This gives a different level of protection for each of high conservation/ecological value systems; slightly to moderately disturbed systems (where the guidelines will mostly be applied); and highly disturbed systems. Hence the approach acknowledges that an area may be degraded and it will not be possible immediately to create a high ecological status. It is of note that this approach mirrors

Table 8.9 Bathing waters standards. Expressed as numbers per 100 ml

Parameter	USA	Europe mandatory[a]	Europe guideline[a]
Total coliforms	700	10,000	500
Fecal coliforms	200	2,000	100
Fecal streptococci	—	300	100
Enteroviruses	—	0	0
Salmonella	—	0	0

[a] Mandatory means that it should not be exceeded, Guideline indicates the desirable aims.

Source: EPA of the United States, and European Community Directives.

that to be taken by the European countries under the EU Water Framework Directive in which most water bodies will be managed to achieve good ecological status, although there will be lower aspirations (termed good ecological potential) for those areas deemed to be Highly Modified Water Bodies or Artificial Water Bodies.

8.3.3 Management of estuaries

It is not possible here to detail the management of every activity occurring within an estuary and so only general principles will be given followed by a few key examples. The management of an activity within the estuary starts with an assessment of likely effects prior to the activity starting or a structure being built. An Environmental Impact Assessment (EIA) should distinguish between substantial, moderate, or slight impact according to: (i) the magnitude of impact itself both spatially and temporally, (ii) the value and sensitivity of receiving area, and (iii) the aesthetic sensitivity. Hence the main steps in performing an EIA are: (1) identifying the potential sources of impact; (2) describing the receiving environment; (3) evaluating the potential impacts; (4) determining the mitigation measures possible, and (5) identifying and quantifying any residual impacts, which may be medium or long term.

Box 8.4 The framework for applying the Australian/New Zealand Water Quality Guidelines (from National Water Quality Management Strategy 2000)

Define
primary management aims
(including environmental values, management goals, and level of protection)
⇓
Determine appropriate
water quality guidelines
(tailored to local environmental conditions)
⇓
Define
water quality objectives
(specific water quality to be achieved)
taking account of social, cultural, political, and economic concerns where necessary
⇓
Establish
monitoring and assessment program
(focused on water quality objectives)
after defining acceptable performance or decision criteria
⇓
Initiate appropriate
management response
(based on attaining or maintaining water quality objectives)

Industries and developers are required to determine the best practicable environmental option (BPEO) to solve the potential or actual environmental problems. The steps in selecting a BPEO are to define the objectives, generate the options, evaluate the options, summarize and present the evaluation, select the preferred option, review the preferred option, and implement and monitor the decision and its effects. For example, an estuarine coastal municipality faced with a city's sewage will use the BPEO process to determine whether the waste should be treated and discharged to sea, incinerated, spread on farmland or forest areas, or put into landfill. The preferred option will be designed to use the best available technology (BAT), although this has been extended to the best available technology not entailing excessive costs (BATNEEC).

As shown in earlier chapters, dredging of navigation channels is a major operation in estuaries and it has the potential to have a major impact on the estuary, hence its management is of central importance. Under national and international legislation and agreements (see below), the quality of the material being dredged has to be analyzed for the presence of pollutants. Highly contaminated material cannot be allowed to be dispersed into the estuarine or adjacent coastal waters and so needs to be contained. While some of this may be put in secure landfill sites, whereby any runoff will not pollute groundwaters, the degree of contamination in dredged material from some estuaries has required large-scale solutions. For example, the Port of Rotterdam on the estuary of the Rhine during the 1990s created two large bunded containment areas—the Papasbiek for the most heavily contaminated dredgings and the Slufter for the less-contaminated material. Both of these were land-claim areas although the former was also sealed by a heavy duty lining to prevent leaching of any pollutants. This solution to the problem had to be accompanied by a reduction in pollution discharged into the Rhine such that in the long term, new dredgings could safely be disposed to sea.

In many cases, dredgings are disposed of into the estuarine waters or out to sea with the management aim being to ensure that the material can be moved by the dredger faster than it is accumulating. More recently, nature conservation bodies have been concerned that such a removal of sediment from an estuarine system will have effects on the natural system, for example, removing sediment from a navigation channel may draw down sediment from intertidal mudflats. Because of this, estuarine managers have advocated either a greater use of estuarine disposal, to keep the material in the estuarine circulation even if this requires a continued high effort by the dredging companies, or the *beneficial use* of the material. An example of the latter is to use dredged material as beach nourishment or recharge onto intertidal areas as a means of countering sea-level rise and beach erosion.

Other activities such as fisheries and aquaculture, mineral exploration and extraction, and recreation are all managed to a lesser or greater extent albeit usually by different organizations and certainly under different legislation and agreements. Where they produce waste or polluting discharges, these are controlled by the relevant pollution abatement legislation and procedures. For example, organic matter produced by intensive aquaculture can be controlled under licences either preventing dumping of pollution into the sea or, as is the case in the United Kingdom, pollution from a land-based source. Aquaculture is subject to management of the persistent chemicals that it uses, such as pesticides, the use of medicines to combat disease and through regulations controlling the use of genetically modified organisms or the introduction of alien species. Fisheries are managed through by-laws, which may control the timing, areas, and methods of fishing and increasingly fisheries legislation is designed to protect the ecosystem from the effects of fisheries, especially the taking of non-target sizes and species of fishes and other organisms.

Mineral exploration and extraction, such as for oil, gas, and aggregates (sand and gravel for construction) is subject to the use of licensed areas, controls on the use of the seabed, and pollution abatement, ecosystem protection and EIA legislation and agreements. Hence, although estuarine areas may be licensed for the extraction of oil and gas, the operations would not be allowed in sensitive areas or at sensitive times of the year, for example, when large numbers of wading birds would be affected by any spillage of oil following the exploration.

Since the late 1980s, perhaps the largest discussion of estuarine management has centered on ways of combating coastline change, sea-level rise and erosion in low lying areas, especially those supporting industry or large populations. Faced with a changing coastline, the estuarine manager has four options: to hold the line, to advance the line, to retreat the line, or to do nothing—the latter is always an option! The first of these is to keep maintaining and raising seawalls to keep pace with rising sea-levels and, as in large parts of North American and Europe subject to isostatic rebound following the last Ice Age, a sinking land mass. The second is to keep building out into aquatic habitats and continue to claim land but with a higher degree of protection. While both of these first two may not be ecologically sustainable, the third option includes the increasingly used setting back of sea defences, previously called managed retreat but now more positively called managed realignment or, in the Low Countries, depolderization. The aim of this is to take previously claimed land or low lying areas and return them to wetland. The latter act as water storage areas in times of flooding, absorbs energy thus reducing erosion, and thus alleviates the need for continually raising the seawalls; in addition it creates wetland habitats, which will offset the removal of these through continuous land-claim. Hence it has been described as a win–win management situation whereby estuarine squeeze is reversed and flood protection is effected. In several areas, such a strategy is also accompanied by building movable barriers to counteract exceptional storm surges and high tides.

8.3.4 Management and protection of catchments and adjacent coastal and marine areas

Throughout this book, it has been emphasized that the functioning of the estuary depends on the well-being of the areas both upstream and at sea. Hence the management of the estuary cannot be separated from the management of the wider areas. Indeed, recent initiatives such as the EU Water Framework Directive will lead to the determination of river basin characteristics and the creation of catchment management plans. Similarly, in the United States there is the management of eco-regions as well as widespread national strategies.

Given the importance of sediment supply to estuaries, it is necessary to manage them in relation to neighboring sediment fluxes, supply, and loss. For example, the coastlines of the United Kingdom have been subjected to Shoreline Management Plans (SMPs) with a focus on the management of coastal protection. A SMP has been created for each sediment cell such that there can be an assessment of coastal protection strategies inside and outside the estuaries. For example, the glacial clay cliffs of the Holderness coast in eastern England are eroding at 2 m per year and thus adding sediment to the nearby Humber estuary. Management initiatives in this area take the view that such an erosion, especially of coastal farmland, should not be prevented otherwise the estuary would be starved of sediment thus creating erosion problems within the Humber estuary. It is of note that these strategies are within a national UK strategy that areas of high urban concentration and industry in the national interest should be protected against coastal erosion whereas low value farmland is a low priority for protection.

8.4 Practice

The management of estuaries has in some cases produced spectacular reversals in the conditions of estuaries after decades, or even centuries, of neglect and pollution. The clean up of the Delaware Estuary represents one of the premier water pollution control success stories in the United States (Albert 1988). The Delaware Estuary, bounded by the states of Delaware, New Jersey, and Pennsylvania, is located in one of the most complex urban industrial regions of the United States (Fig. 8.1), having one of the world's greatest concentrations of heavy industry, and America's second

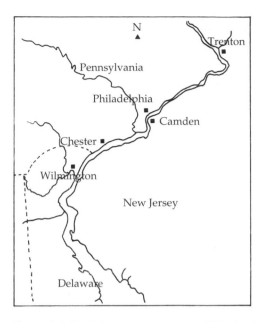

Figure 8.1 The Delaware estuary, eastern USA, showing the main centers of population. (After Albert 1988.)

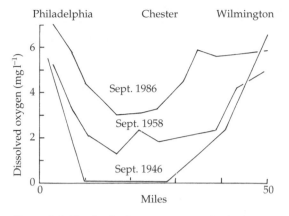

Figure 8.2 The dissolved oxygen concentration in the Delaware estuary, in 1946, 1958, and 1986. Measurements were made under similar conditions of tide, temperature, and freshwater discharge. For locations see Fig. 8.1. (After Albert 1988.)

busiest port, and second largest complex of oil-refining and petrochemical plants. A population of over 6 million people resides in the region. Once largely devoid of aerobic aquatic life, now, thanks to a long-term improvement program, it supports a variety of recreational uses, year-round fish populations and a number of key migratory fish species.

Improving the quality of the Delaware Estuary has been an evolutionary process spanning almost 200 years, which is summarized in Table 8.10. When Europeans reached the Delaware in 1609 it was presumably pristine, but by 1799, when the first pollution survey was undertaken, a variety of pollution sources were noted, and in response to pollution of drinking water supplies, the first municipal water supply and sewer systems were constructed. As population, and disease grew, improved water supply and sewage systems were installed, which solved public health concerns, but did nothing about the pollution of the estuary. The greatest degradation of the estuary occurred in the first

half of the twentieth century, following expansion of industry and population. From 1914 to 1937 surveys indicated that the estuary was substantially polluted, with oxygen levels only just above zero. By 1946 an area of anoxia had developed running shore-to-shore and down the estuary for 20 miles (35 km) (Fig. 8.2). Remedial action undertaken thereafter, involving the clean up of rivers and streams entering the estuary, substantial reduction in organic discharges to the estuary and dredging of coal silt was undertaken, all with the aim of restoring dissolved oxygen levels to at least 50% saturation. As a result of these activities, the estuary became no longer anoxic, but it was still experiencing depressed dissolved oxygen levels, indicating continuing organic pollution (Fig. 8.2). In the early 1960s new emission standards were adopted, which required the upgrading of most sewage-works from having primary treatment, to full secondary treatment. As a result the dissolved oxygen concentration has risen steadily (Figs 8.2 and 8.3), and the fecal coliform levels have dropped (Fig. 8.4). There is still a seasonal sag in the oxygen content of the water (Fig. 8.3), but as a result of these improvements the

Table 8.10 Description of the five generations of Delaware Estuary water pollution control efforts. From Albert (1988)

Generation	Approximate time span	Problem	Actions	Prime participants[a]
First	1800–60	Pollution of local water sources	Construction of municipal water systems with river intakes, some sewer line construction	Municipal government
Second	1880–1910	Water borne disease from consumption of river water	Construction of water filtration plants, development of alternative water supplies, sanitary sewer system construction	Municipal government
Third	1936–60	Gross pollution	Construction of primary wastewater treatment plants after effluent standards adopted	INCODEL, states
Fourth	1960–80	Substantial pollution	Construction of secondary or higher wastewater treatment plants after wasteloads allocated	DRBC, states, and USEPA
Fifth	1980–present	Public health and aquatic life concerns, including toxins	Underway. Includes combined sewer correction, more stringent point source controls, non-point source controls, toxic materials controls and others	DRBC, states, and USEPA

[a] The efforts of the federal government prior to the creation of the United States Environmental Protection Agency (USEPA) and the efforts of most cities and industries to clean up their wastes in the third and fourth generation effort are recognized as well. INCODEL, Interstate Commission on the Delaware River Basin. ORBC, Delaware River Basin Commission.

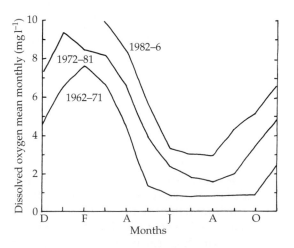

Figure 8.3 Mean monthly dissolved oxygen concentration in the Delaware estuary at Chester, averaged for each of three decades. Note the decrease in the duration of the critical low dissolved oxygen period with time. (After Albert 1988.)

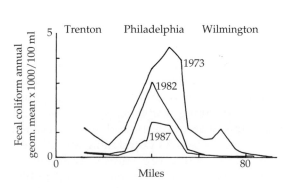

Figure 8.4 Annual geometric mean count of fecal coliform bacteria in the Delaware estuary for 1973, 1982, and 1987. (After Albert 1988.)

diversity of fish increased from 16 resident species in the 1960s, to 36 by 1985. Improvements are still required, both to dissolved oxygen levels, and to the control of specific toxic substances, but the Delaware Estuary has been established as a prime example of how mankind can restore or repair the damage, which he has caused to the estuarine ecosystem.

The Thames Estuary, UK, with the city of London and approximately 10 million people on its banks, has experienced similar problems to the Delaware, and like it has now been dramatically restored. On the Thames, there was a steady decline in the oxygenation of the estuary from 1920 until 1955, so that periods of total deoxygenation regularly occurred. Since that date improvements, mainly to sewage-works, have resulted in a substantial improvement in the condition of the middle reaches. The recovery of the diversity of fish species in the Thames, as was shown in Fig. 6.9, amply shows the success of the management program.

Fortunately not all polluted estuaries are, or have been, as badly polluted as the Delaware or the Thames, but it is also true to say that few have been as successfully treated as those two. The lesson of the Delaware and the Thames is that major improvements to sewage-works and improved treatment of industrial effluents can restore the oxygenation state of an estuary, so that the environmental quality objectives can be attained. In both cases the restoration of the estuary has taken decades of effort, and large sums of money. These recovery programs have also clearly shown that the further downstream the discharges can be made, then the lesser will be the resultant pollution, since there will be a larger body of constantly changing water to receive, dilute, and degrade the pollutants. In both of these estuaries the discharges to the estuary have been reduced to levels that were within the assimilation capacity of the estuary, so that the normal functioning of the estuarine ecosystem could cope with the material entering it.

Both the Delaware and the Thames were extreme problems of organic enrichment in estuaries, which has been shown to be reversible in its effects. However, the problems of toxic chemicals, especially those accumulative substances on the black lists, may be less readily reversible, and will continue to pose problems to estuarine managers. The recovery of the ecosystem is largely dependent on the processes of sedimentation. Thus the concentration of the pollutant in the water column may decline rapidly after the cessation of discharge, but its concentration within the sediment may continue to influence the fauna and flora inhabiting the sediment. Any re-suspension of the sediment, due, for example, to dredging, will serve to liberate the pollutant into the water again. The intertidal and other habitats of an estuary can utilize (or bury) many of the pollutants that human activity introduces into estuaries, but such utilization (or burial) depends on the maintenance of all aspects of a healthy functioning ecosystem.

8.4.1 Monitoring for management

As an integral part of management practice, monitoring should occur in order to detect change and to determine whether the management is having a desired effect. For example, monitoring strategies in the Humber estuary, Eastern England, are designed to reflect pollution abatement schemes by industry. As an integrated program, the Humber bioaccumulation monitoring program measures the levels of trace metals in:

- a seaweed (*Fucus vesiculosus*) to reflect levels in solution
- an infaunal polychaete, the ragworm (*Nereis diversicolor*) to indicate the transfer of pollutants from a sink such as the sediments into a primary food source for the higher predators
- an epifaunal crustacean, the brown shrimp (*Crangon crangon*), a major food source of fishes in the estuary
- an estuarine resident fish, the flounder (*Pleuronectes flesus*), to reflect uptake from sediments and infauna

• and a marine, commercially taken fish, the Dover sole (*Solea solea*).

In addition, the sources of such pollutants are measured in the water column, sediments, riverine inputs, industrial discharges, and domestic discharges. This monitoring is complemented by monitoring of the intertidal and subtidal invertebrate communities, shrimp populations, resident fish community and migratory salmonids, the water quality, inputs and ambient levels of persistent organics, hydrocarbons and nutrients, flow characteristics, and effluent loadings.

In addition to the above type of monitoring, most major estuaries in North-west Europe are also subjected to a harmonized monitoring scheme. This is under the auspices of the Paris Commission and is designed to produce an inventory of materials entering the European shelf sea areas.

8.4.2 Legislative—types of instruments

National

Before managing an area, it is necessary to define the area of competence, that is, who is the competent body responsible for implementing legislation (such as the Environment Agency, EA in England and Wales) and to define the area to be controlled (hence "controlled waters" under COPA). In the United Kingdom, since its implementation in 1985, the dominant piece of anti-pollution legislation is the Control of Pollution Act 1974: Part II Pollution of Water (COPA II). This is aimed specifically at "poisonous matter and solid wastes" and "trade and sewage effluent" to rivers, estuaries, and other water courses. Its main provisions set down the offence of pollution into relevant, specified, and controlled waters, it controls Trade Effluents, deals with applications for consent to discharge; ensures publicity and freedom of information; and is enabling legislation for EC Directives and OSPARCOM. It has been updated and revised by the Water Act 1989, Environmental Protection Act 1990, and Water

Resources Act 1991; it extends IPC, BATNEEC—replaces to some extent BPEO and Best Practical Means, using the EQO/EQS approach; note that harbor authorities also have power to establish by-laws to regulate dumping of polluting substances under Rivers (Prevention of Poll.) Acts 1951, 1961.

EU legislation

The largest source of environmental legislation for European countries is the EU and in many ways the environmental direction mirrors that of other developed countries such as the United States and Australia. There are common themes throughout the environmental legislation of these large groupings of member states and so the EU is given here as an example of environmental thinking amongst a large block of autonomous states.

The EU environmental policy is mainly through *Regulations* (which are directly applicable as law and are generally used to achieve precise goals) and *Directives* (which establish the goal and results to be achieved but allow flexibility and leave the method of implementation to the state); thus the latter give some measure of discretion by each state (this is termed "subsidiarity"). A Directive is applicable in all member states (or those to which it was issued) and members must submit to the EU Commission the national laws implemented to meet directives (this is called their enabling legislation). The member state is required to issue "letters of compliance" detailing steps to incorporate the directive into its national legislation. If the performance of a member state is unsatisfactory then it is reported to the European Court of Justice and although member states can be stricter in environmental regulation than the EU they cannot be less strict. Even within this short description, North American readers will see the close similarities between the EU and the federal system in the United States.

Tables 8.11 and 8.12 summarize the EU Directives of relevance to estuaries. Many of the earlier directives had a sectoral basis, that

Table 8.11 Relevant EU Sectoral Directives and their impact on Estuaries

Directive	Summary of main features	Impact on estuaries
Dangerous Substances + Daughters 76/464/EEC etc.	A Framework Directive, provides for a system of licensing by consent before discharge of listed substances; allows for Uniform Emission Standards (UES) and EQS Approaches; basis of List 1 and List 2 compounds and derivatives. The directive requires an inventory of sources and probable sources from users and sewers (data required on: max. conc. ($\mu g\,l^{-1}$), max. flow. ($m^3\,day^{-1}$), max. Monthly load (kg), estimated load discharged (kg year^{-1}), plant handling capacity	Reduction of inputs to catchments, control on use of polluting materials, prioritization of the most polluting materials; this links to the IPPC and Water Framework Directives (see Table 8.3) and will eventually be replaced
Urban Waste-water Treatment 91/271/EEC	Provision of sewage treatment to minimum standards, based on the population equivalent (PE), as a fixed amount of organic matter from the population and as a means of converting organic industrial discharges to a PE and type of receiving area	Changes to estuarine discharges; often long sea outfalls following only preliminary screening from STW for larger urban areas
Integrated Pollution Prevention and Control 96/61/EC	Basis in controlling waste from prescribed industrial processes to all media (land, water, and air) and acknowledges a transfer of pollutants across media; aim to ensure that industries bring in BAT to control waste	Ensures that all major chemical plants alongside estuaries have Environmental Management Systems and control pollutants from all sources
Titanium Dioxide 78/176/EEC	A single industry relevant Directive, designed to reduce polluting discharges, to ensure a licensing system, inventory of discharges and monitoring system	The largest European TiO$_2$ plants are on estuaries such as the Humber and Tees estuaries in the United Kingdom
Shellfish Growing Waters and Health 79/923/EEC	Designed to monitor and improve waters in which shellfish grow, ensures monitoring of water quality, especially aesthetic and polluting aspects, in areas suitable for shellfish (to be repealed by the Water Framework Directive)	This ensures that areas are designated within estuaries and coastal water bodies and then any polluting discharges within those areas are treated
Shellfish Harvesting 91/942/EEC	Areas designated as Class A (shellfish can be taken and sold for consumption without treatment), Class B (shellfish require placing in purification waters (recirculating system) before being suitable for selling), Class C (shellfish require re-laying for a given period in order to improve quality), Class D (shellfish too contaminated or waters too polluted to allow consumption)	Has led to areas designated in many estuarine areas, for example, Morecambe Bay in the United Kingdom, where mussels are taken
Quality of Bathing Waters 76/160/EEC and revisions	Set quality standards for "recognized bathing areas"; set out frequency of sampling and procedures to be used; define circumstances in which directive may be waived; require reporting at regular intervals. Parameters: total coliforms, faecal coliforms, enteroviruses, pH, transparency, *Salmonella*, color, mineral oils, surface-active substances (foam), phenols, odors	Ensures that estuarine beaches are sampled on 20 occasions during the bathing season (e.g. May–September) and any breaches are tackled by improvements in sewage discharges
Quality of Freshwaters to support Fishes 78/659/EEC, amend 91/662/EEC	Requires member states to designate waters, which need protection or improvement in order to support fishes; it is primarily concerned with salmonid waters and cyprinid (coarse fish) waters; (to be repealed by the Water Framework Directive)	Of minor interest to estuaries except that it is designed to protect the waters receiving or delivering Diadromous fishes from or into estuaries
Deliberate release and use of GMO 90/219/EEC, 90/220/EEC repealed by 2001/18/EC	Controls the use of genetically modified microorganisms and the deliberate release into the environment of genetically modified organisms; the latter including those used in aquaculture	Aquaculture in estuaries and other transitional waters using genetically modified salmonids and shellfish will be controlled

Table 8.12 Relevant EU Holistic/Ecosystem Directives and their impact on Estuaries

Directive	Summary of main features	Impact on estuaries
Environmental Impact Assessment (85/337/EEC and 97/11/EC)	On the assessment of the effects of certain public and private projects on the environment—EIA to identify, describe, and assess in an appropriate manner the direct and indirect effects of a project on the following: (i) humans, fauna, and flora; (ii) soil, water, air, climate, and the landscape; the interaction between (i) and (ii); material assets and the cultural heritage	All major estuarine developments will be subject to EIA and local planning regulations. This will require adequate studies, mitigation, and compensation measures
Conservation of natural habitats and of wild fauna and flora (termed Habitats Directive) 92/43/EEC	Basis in habitat protection by creating a network of designated sites (Special Areas of Conservation, SAC) throughout Europe; SACs were to be proposed by 1995 but by early 2000s sites were still being proposed and adopted as SACs. Annexes set out why and where SAC to be designated—Annex I gives the priority habitats and Annex II gives the species of concern; sites are to have schemes and committees of management; enabling legislation in Member States, competent authorities designated to use existing functions, relevant authorities to use their powers, but cannot interfere with legitimate users and uses	Annex 1 habitats include estuaries, intertidal sand and mudflats and subtidal sandbanks, salt marsh and sea grass beds amongst others. Annex 2 species include several estuarine fish species: sea and river lamprey, shads, salmon, and sturgeon
Conservation of Wild Birds 79/409/EEC	Member States are required to take measures including the designation of Special Protected Areas; designation to maintain sufficient habitat for the support and protection of all bird species, also conservation measures for rare, vulnerable, and migratory species. SPA will be combined with SAC to give the *Natura 2000* network of sites across Europe	Estuaries supporting internationally important populations of wading birds have been designated as SPA
Nitrates—level, control, use 91/676/EEC	With relevance mainly to agricultural sources of nitrogen compounds, includes controls on nitrate use and runoff, aims to combat actual and potential eutrophication and health aspects due to excess nitrates in water courses and groundwater; means of tackling diffuse nutrient inputs in order to ensure surface waters	The aim is to prevent estuaries from developing symptoms of eutrophication—especially macroalgal mats, toxic and nuisance algal blooms
Strategic Environmental Assessment 2001/42/EC	Adopted to provide for a high level of protection of the environment and to contribute to the integration of environmental considerations into the preparation and adoption of plans and with a view to promoting sustainable development	Will ensure a more integrated assessment of environmental consequences due to estuarine developments
Water Framework Directive 2000/60/EC	Main Objectives of Sustainable Water policy: provision of drinking water, provision of water for other economic requirements, protection of the environment, alleviation of the impacts of floods and droughts, will provide overall framework for European, national and regional integrated and coherent water policies. (This Directive will repeal several others)	Resulting in definitions of types of estuaries, called Transitional Waters in the Directive, classification of quality, reference conditions against which other estuaries can be compared, and the determination of good ecological status for estuaries
Environmental Liability for preventing and remedying	Aim: to address effectively and efficiently site contamination and the loss of biodiversity, to establish a framework whereby environmental damage would be prevented or remedied. Mode of achieving this to be	To Link with the Habitats and Water Framework Directives and related to biodiversity and human health

Table 8.12 (*Continued*)

Directive	Summary of main features	Impact on estuaries
environmental damage (COM(2002 17 Final) proposed Directive)	left to Members States under subsidiarity and complementarity principles. It aims at prevention of damage; based on polluter pays principle. It will cover occupational activities posing a risk, requiring regulatory requirements and aims to stimulate prevention, restoration and cost-recovery (under polluter pays principle)	
Future?	The EU has produced 2 Communications (1) on Integrated Coastal Zone Management: a strategy for Europe (COM (2000) 547 Final) and (2) Towards a Strategy to Protect and Conserve the Marine Environment (COM (2002) 539 Final). These may be translated into Directives and will aim to integrate many of the Directives mentioned here	These strategies enable estuarine management to be integrated with all other coastal and marine systems

is, they were designed to tackle a particular problem such as the quality of bathing beaches or the discharge for a specific industry such as titanium dioxide production (Table 8.11). Recent years have, however, shown a movement to a greater holistic, integrated, and ecosystem approach with very wide ranging directives (Table 8.12). In addition, community-wide agreements such as the Common Agriculture and Common Fisheries Policies now have a wider environmental protection role. For example, the Common Fisheries Policy now includes the protection of marine habitats and species in the management of fish stocks.

Within Europe, the Commission of the European Union has been particularly active in issuing directives for the control of pollution, based on the UES approach, and has in recent years issued a series of standards through their directives. The directives are community law, but each needs a nation's own law to enable them to be adopted, and the nation has the choice of using the UES or EQS approach. In addition there are other wide-ranging EU Agreements, which affect estuaries such as the Integrated Coastal Zone Management (ICZM) strategy. The Marine Strategy published in 2003 will integrate many of the marine environmental aspects given through EU policy and legislation.

International obligations

Most countries have agreed, as signatories, to wider international and in many cases, global declarations. These include:

The UN Convention on Environment and Development (UNCED, the so-called Earth Summits) in 1992 in Rio de Janeiro, in 1995 in New York and 2002 in Johannesburg. These laid down the general principles for ensuring sustainable ecosystems and the 1992 convention led to Agenda 21, which aimed to set the philosophy for the twenty-first century by encouraging states and individuals to "think global, act local." The UNCED led to the Convention on Biological Diversity (CBD), which provides a legal framework for biodiversity conservation through the development of national action plans to halt the worldwide loss of animal and plant species.

The International Maritime Organization (IMO), especially through its MARPOL agreements, aims to control litter and pollution in seas, the use of antifouling paints on vessels and the introduction of alien organisms in ballast waters.

The UN through its Convention on the Law of the Sea (UNCLOS) also lays down general principles on the management of the marine areas and the prevention of harm to those systems. Amongst others, the UN Food and Agricultural Organization has adopted and promoted several underlying legal principles to support (ICAM) Integrated Coastal Area Management: the precautionary principle, the principle of preventative action, the polluter pays principle, the responsibility not to cause trans-boundary environmental damage, the rational and equitable use of natural resources and the involvement of the public (www.fao.org/docrep/w8440e00.htm).

Many countries are also signatories to regional environmental conventions and are thus members of commissions. These include the Oslo and Paris Commission for the NE Atlantic including the North Sea (OSPAR, now combined into the 1992 Paris Commission), the Helsinki Commission for the protection of the Baltic (HELCOM) and the Barcelona Convention for the protection of the Mediterranean (MEDPOL). Although each of these originally was formed to prevent pollution both from land and vessels, their remit has been extended since the 1990s to include an ecosystem approach. These conventions have produced harmonized monitoring schemes to measure pollutants coming through their estuaries and more recently have produced Action Plans and Quality Status Reports of their areas. OSPAR, notably, has produced strategies to combat eutrophication and pollution by hazardous substances within its estuaries and coastal regions. Similarly, sea areas are likely to have groupings of adjoining states in order to control activities in them, their estuaries and catchments. For example,

the North Sea Ministerial Conferences have taken place approximately every 5 years since the early 1980s. These have led their member states to agree to estuarine and coastal management measures to combat environmental deterioration and excessive use. However, there is some discussion over the value of these commissions in areas where they overlap with other collections of member states such as the EU (see Ducrotoy and Elliott 1997).

As well as being signatories to regional agreements, countries will agree to global conventions. For example, all developed states have a legislation and administrative system to control the dumping of wastes into the marine or estuarine waters. While in Europe, the dumping of industrial waste was banned in 1992 and sewage sludge in 1998, the disposal of dredged material requires a licence. Those countries are signatories to the global London Dumping Convention (now London Convention) on the Prevention of Marine Pollution by Dumping of Wastes and Other Matter. Hence they need enabling legislation to adopt agreements under that convention. Part IV of Canada's Environmental Protection Act 1988, the Republic of Ireland's 1981 Dumping at Sea Act, and the United Kingdom's 1985 Food and Environmental Protection Act (FEPA), for example, provide for the licensing and control of materials dumped into estuarine and coastal waters.

8.4.3 Administrative/organizational management framework

Given the plethora of legislation illustrated above, it is necessary to consider which administrative bodies are responsible for implementing it. The estuarine management organizations will implement a framework based on national and international legislation, agreements and protocols such as those illustrated above. This in turn has created, in most countries, an excessively complex administrative and organizational structure for managing estuaries. The Box 8.5 illustrates

---

Box 8.5 Case study—Estuarine and coastal management framework in Scotland (2003), as an example of the legal complexities of estuarine management. Similar situation in other countries

---

Planning and Development
National Planning Policy Guidelines
Planning Advice Notes
Structure (General or Strategic) Plans
Local (Detailed) Plans
Civic Government (Scotland) Act 1982
Water Environment & Water Services
  (Scotland) Act 2003
Town and Country Planning (Scotland) Act 1997
Environmental Assessment (Scotland) Amendment
  Regulations 1997
FEPA

Landscape
National Scenic Areas (NSA, SO Env. Dept. Circular)
Area of Great Landscape Value (AGLV)
National Parks (Scotland) Act 2000

NGO site designations
RSPB Reserves
SWT Reserves, Voluntary Nature Reserves

Fisheries
EU Common Fisheries Policy (quotas,
  conservation, fleet structure)
Scottish Office Regulatory Orders
SO/SNH Code of Guidance (e.g. cockles)
Inshore Fishing (Scotland) Act 1984 (statutory
  instruments)
Sea Fisheries (Shellfish) Amendment (Scotland)
  Act 2000
Several & Regulatory Orders
Inshore Fishing (Prohibition of Fishing and Fishing
  Methods) (Scotland) Order 1989
Salmon and Freshwater Fisheries (Protection)
  (Scotland) Act 1951
Sea Fisheries (Conservation) Act 1967 (licences,
  restrictions, landing sizes)

Nature conservation
Wildlife & Countryside Act 1981 (Amendment)
  (Scotland)
Regulations 2001 (Section 28 - SSSI)
Designation of PDO (Potentially Damaging
  Operations [by SNH])
EU Wild Birds Directive (for SPA)
EU Habitats & Species Directive (for SAC)
The Conservation (Natural Habitats, &c.)
  Regulations 1994
Scottish Office Environment Department Circulars
Nature Conservation (Scotland) Bill (draft)

Combined designations
Environmentally Sensitive Areas (ESA, 1986
  Agriculture Act)
Natural Heritage Areas (NHA, Natural Heritage
  (Scotland) Act 1981)
UK Action Plan for Biodiversity 1994 (from UNCED)

Coastal protection
Scottish Office Policy
Local Authority protection & regulation
Food & Environment Protection Act 1985
  (below HW)
Coast Protection Act 1949
Flood Prevention and Land Drainage (Scotland)
  Act 1997
Environment Act 1995
National Planning Policy Guidance

Aquaculture
Disease of Fish Act 1937 (Scottish Office control)
Crown Estate Act 1961 (lease of area)
Department of Transport (installation, hazards to
  navigation)
Water quality regulations (SEPA, EU)
Planning Authorities (control above HW)
Health & Safety Executive (safe operations)
Environmental Assessment (Salmon Farming in
  Marine Waters) Regulations 1988
Town and County Planning (Scotland) Act 1997
  (development control)
Water Environment and Water Services (Scotland)
  Act 2003
Wildlife & Countryside Act 1981 (pest control)

Box 8.5 *(Continued)*

Industrial development
National Planning Guidelines
Petroleum (Production) Act 1934 (Crown
  ownership of oil & gas)
Licensing rounds (Dept. of Trade & Industry)
Electricity Act 1989 (permission for proposal, SoS)
EU Directives, for example, EIA Directive
Food & Environmental Protection Act 1985
  (SERAD, for example dredging)

Pollution control
Environment Act 1995 (creation of SEPA)
Control of Pollution Act 1974 (and predecessors
  and successors)
EU Bathing Waters Directive & UWWTD
EU Dangerous Substances Directive (and daughters)
Food & Environmental Protection Act 1985
  (sewage sludge, dredging)
1973 Convention for Prevention of Pollution from
  Ships (MARPOL, Annex V)
Merchant Shipping (Protection of Pollution by
  Garbage) Regulations 1988
Environmental Protection Act (1990) (litter, landfill sites)
OSPARCOM 1974/1992
Trade effluent controls (sewerage undertakings)
Pollution Prevention and Control (Scotland)
  Regulations 2000
Water Environment and Water Services (Scotland)
  Act 2003

Overall
UN Convention of Environment & Development
North Sea Ministerial Conferences (1995, 2002)

this complexity for one country, Scotland, but it is emphasized that this feature is encountered in many others (e.g. Fernandes *et al.* 1995). Although this structure in many countries changes often, and usually following government changes, the structure gives the main instruments for controlling and managing human activities.

The integrated management of estuaries is included in ICZM, which relies on hierarchical organization, that is, where different stakeholders and players link with each other. This relies firstly on vertical coordination, which is aimed at creating cooperation amongst various hierarchical levels of government and administration—at national and central, regional and local. Second, it relies on horizontal coordination—creating cooperation within a specific level of hierarchy, for example, at the local level among and across local government, all sectors and various stakeholders, each of which have no responsibilities for the

other stakeholders but which have to link together and cooperate in order to bring about sustainable management. This system also relies on temporal and spatial coordination—the aim to achieve the optimal phasing of management actions across places within the estuary and at different times (see also www.fao.org/docrep/w8440e00.htm).

Coastal Zone Management has been defined as "A dynamic process in which a coordinated strategy is developed and implemented for the allocation of environmental, socio-cultural, and institutional resources to achieve the conservation and multiple use of the coastal zone." (Gubbay 1990). Similarly, "Coastal Zone Management typically is concerned with resolving conflicts among many coastal uses and determining the most appropriate use of coastal resources." (Sorensen *et al.* 1984). In the United States the Environmental Protection Agency (EPA), and the Coastal Zone Management Act (CZMA) have been the main federal instruments for the protection of coastline environments. Since 1972, all 35 coastal states have participated in the CZMA program. Each individual state has produced legislation for its own shoreline, and inevitably it has proved hard to maintain Federal consistency, so that standards for estuarine management do vary from state to state. Each state also has a State EPA together with county environmental regulations such that a developer may need permits from several levels in a hierarchy.

Global and supranational

The international agreements indicated above are usually created by a Conference or Convention attended by states and then leading to a more permanent Commission, the latter having its own Secretariat. The final declaration, protocol, or agreement is then applicable to its signatories although the success of the implementation of these relies on member states bringing in enabling legislation. In addition to this, states are increasingly part of larger, supranational blocks such as the EU and so it is valuable to consider the nature of the latter as an example.

It is arguable that the EU is the largest driver of environmental protection for the countries which are either members, wish to become members, or by choice follow the EU policies. The Environment and Fisheries Directorates, part of the Commission, together with the relevant Councils of Ministers, which are drawn from each member state, and the Parliament, create and agree the EU statutes, principally Directives and Regulations. The Commission will also create and implement EU policies and initiatives (e.g. environment action programs). The European Court of Justice, the judiciary, then will interpret the meanings of treaties and other legislation and decide whether states are adhering to letter and/or spirit of the legislation. In addition to this, the European Environment Agency (EEA), which is not a body of the EU, plays a role in helping to frame EU environmental policies as well as collate and disseminate environmental information.

National

Estuaries are the sites of many activities, uses and users, each of which requires to be managed and consequently, this requires many organizations to be involved. See Box 8.6 as an example of one English estuary, the Humber.

In many countries, the overall environmental control machinery starts with high level parliamentary of advisory bodies. For example in the UK, the Parliamentary Select Committee on the Environment, Food and Rural Affairs will scrutinize Government decisions and policy, and inform and respond to public debate. In turn, the Government Department of the Environment, Food, and Rural Affairs (DEFRA) is a central body with overall responsibility for environmental matters within England— mainly policy but including discretionary, political decisions including water policy. There is other involvement by the Department of Trade and Industry (DTI), Department of Transport, Local Government and the Regions (DTLR), and the Office of the Deputy Prime Minister (ODPM). These departments have the power to approve actions of regulatory bodies and make appointments to regulatory bodies. DEFRA control the funding of regulatory bodies, including the Environment Agency, and has responsibility for the conservation agencies whereas, perversely, DTLR and ODPM controls aggregate extraction and offshore windpower. The devolved Governments in Scotland and Northern Ireland have created the Scottish Executive Rural Affairs Department (SERAD) and in Northern Ireland Environment and Heritage Service and Department of Agriculture and Rural Development (DARDNI). These government departments oversee the respective environment protection agencies and statutory nature conservation bodies of the countries of the United Kingdom. This complex structure has been criticized for creating fragmentation, duplication, and confusion in environmental control (e.g. Carter 1988).

Box 8.6 Case Study: Organizations involved in the management of one estuary, The Humber Estuary, Eastern England

- *Unitary and Municipal Authorities*—countryside management, local plans, economic development, emergency planning
- *Regional Development Agencies*—Regional development funding—EU RDF grants
- *English Nature*—nature promotion and conservation, management of SSSIs, SPA, and SAC
- *Countryside Commission*—land management, conservation, and recreation management
- *DEFRA*—fisheries management, disposal of wastes to sea, flood defence, coastal protection
- *DTLR*—town and country management
- *ODPM*—aggregate extraction
- *North Eastern Sea Fisheries Committee*—management and conservation of shell- and fin-fisheries
- *Environment Agency*—environmental quality of inshore waters, flood defence, sea defence and tidal barriers, conservation and fisheries, integrated aerial, water, and land waste management, health and safety
- *Internal Drainage Boards*—maintenance and operation of sluices

- *Crown Estates Commissioners*—management of foreshore and seabed (obtain revenue from it; ownership)
- *Associated British Ports*—vessel control and navigation maintenance, pilotage, oil-spill contingency
- *Private wharves*—navigation, dredging
- *Sports Council (Yorks. & Humberside Region)*—promotion of recreation
- *Water Services plc (Anglian, Yorkshire, Severn-Trent)*—water supply and sewerage, treatment, and disposal of sewage
- *Ministry of Defence* (RAF)—military use areas;
- *Rail Authority*—maintenance of foreshore with rail connections
- *Voluntary conservation bodies* (RSPB, YWT, LTNC, BASC, etc.)—promotion, conservation, and management of certain areas for own interests
- *Humber Forum*—assists liaison, promotes contact, encourages, and advises for economic development
- *Others*—industry, farmers, landowners, Sailing clubs—maintenance of creek navigation

---

The United Kingdom also has a Royal Commission on Environmental Pollution (RCEP) which is an independent body with a nominated and appointed membership, which meets on an *ad hoc* basis to consider and advise on specific topics. The RCEP started in 1970 but previous Royal Commissions have met since late 1800s. The 1972 Report of the RCEP considered Estuary Quality and included the first discussion of EQO while other annual reports include Tackling Pollution—Experiences and Prospects (1984) and BPEO (1988). It is of note that the EQO adopted for the Humber Estuary, Eastern England, were based on those proposed by Royal Commission on Environmental Pollution, third report 1972.

Many countries now have EPAs, for example, in the United States there are state and Federal EPA, Canada and Sweden have EPA, and in the United Kingdom, each country has an environment protection body. In Australia, each state or territory has an Environment Protection Authority or Agency or Department. In the case of England and Wales, the Environment Agency is an independent corporate body, which is managed by an executive board nominated by the Secretary of State and funded by direct grant and a charging scheme. Its main roles are in water resources protection and management, flood defence, fisheries and navigation, pollution control-waste management licensing, control over transport of waste, IPC processes, and the investigation and clean up of contaminated land. It aims to promote conservation and the enhancement of amenity, and to assess and prevent, minimize, remedy, or mitigate the effects of pollution. Most notably, and in contrast to its forerunners, it has a cost–benefit duty, that is to consider the costs and wider societal benefits when seeking to achieve its general duties.

The UK environment agencies, under the Polluter Pays Principle and based on the Water Act 1989 and it successors, can set a charging scheme in which each discharger is to pay an annual charge for attendance (cost of visits, sampling, irrespective of distance), compliance monitoring (analysis and reporting, reliant on complexity and determinands in effluent), environmental monitoring of the receiving waters, and for the application and issuing of the licence. The level of monitoring is decided by the agency and so greater costs are levied when there is a more complex waste stream, the greater effort required in setting a licence and in monitoring compliance. As an extension of this Principle, monitoring is increasingly being carried out by and directly paid for by the discharger. Hence, it becomes economically viable to minimize waste dis charged into the nearby surface waters.

The UK environment protection agencies (Environment Agency in England and Wales, Scottish Environmental Protection Agency (SEPA) and EHS in Northern Ireland) are responsible for issuing consents for individual land-based discharges to air, land, and water. Nature conservation is the statutory remit of the country agencies—English Nature, Scottish Natural Heritage, Countryside Council for Wales and Environment and Heritage Service in Northern Ireland, who are coordinated through the Joint Nature Conservation Committee (JNCC). These bodies declare estuarine sites as nature reserves, or sites of special scientific interest (SSSIs), and manage the European designations—SPA and SAC.

In the Netherlands, a strong government policy on estuarine and coastal zone management is in operation, implemented by the Rijkswaterstaat, especially the Tidal Waters Division. Similarly, Australia and New Zealand have collaborated to produce water quality guidelines via the National Water Quality Management Strategy in Australia and the New Zealand Resource Management Act. These countries have moved toward an adoption of best practice and cooperative best management. This has required a change from control to prevention, from focusing on prescriptive regulation to that on outcomes, and an emphasis on cooperation rather than direction. Their Water Quality Guidelines encourage cooperation between all stakeholders to maintain or improve water quality. Hence they use tools such as the creation of memoranda of understanding, impact assessment, catchment management plans, and monitoring.

Local
Given the plethora of organizations involved in management of estuaries, there is the need to achieve coordination. For example, regional and municipal bodies are often charged with coordinating the many bodies involved with an estuary as well as the need to conserve and preserve the environment. They may use structure plans to foster environmentally sustainable development of natural and man-made resources and economic regeneration, but to take into account sea-level rise, and to minimize adverse effects on the environment. By their nature, the structure plans have to consider water quality, coastal flooding, undeveloped coasts, natural and cultural heritage, special habitats, and hazardous industry.

As an example, coastal and estuarine regional municipalities and local authorities in the United Kingdom have a role to consider and refine Structure Plan policies for their coastlines. For example, within the Humber estuary, Eastern England, the municipalities have adopted policies in line with managing a low lying area subject to increasing sea-level rise and isostatic rebound (i.e. land sinking since the last Ice Age). There is a presumption against new development in areas liable to flooding or in areas which may require an enhanced level of protection, and they have to ensure that the development of land bordering the estuary does not result in the loss of irreplaceable environmental resources such as mudflats. The body should ensure that the demand for coastal/estuary protection does not lead to the need for flood defences, which would have a detrimental effect on estuarial environments. This is against a background of a governmental strategy for flood defence and coastal erosion, that discourages building in areas susceptible to flooding/erosion, and an Environment Agency strategy to protect existing facilities with coastal defences but not to impede the natural hydrography.

NonGovernmental Organizations

In most countries, nongovernmental organizations (NGOs) play a role in public awareness and are often a non-statutory consultee in developments. For example in the United Kingdom, there are many strong conservation-based NGOs such as the Royal Society for the Protection of Birds (RSPB), the Wildlife Trusts, the Worldwide Fund for Nature (WWF), the Campaign for the Protection of Rural England (CPRE), Greenpeace and Friends of the Earth. These bodies may promote nature conservation and natural resources, monitor and comment on local and structure plans, provide advice on conservation issues, purchase or lease land for nature reserves, monitor conservation value and promote the need for further research and cooperative effort.

These bodies in general have an underlying theme of working toward sustainable development, that is, the use of natural resources in order to meet existing demands but not to jeopardize future uses.

Estuarine classification schemes and standards/objectives: Deviation from normality

A central theme in the management of estuaries is the use of standards and objectives, which indicate what is the normal situation and to measure departure from that normality and then to bring in management options to restore normality. This occurs throughout many policies and instruments and may have different names—for example, defining Reference Conditions within the EU Water Framework Directive, Favourable Conservation Status within the EU Habitats Directive, Ecological Quality Objectives (EcoQOs/EcoQS) as defined by OSPAR, and Reference sites under the Australian/NZ Water Quality Guidelines.

An estuary or its parts are often classified by managers in order to (a) describe the overall quality of the system, (b) indicate which areas need action to be improved, restored, or rehabilitated, or (c) indicate the effectiveness of any actions taken. While there are many such classification schemes in operation worldwide, the example adopted by the UK environment protection agencies (Box 8.7) gives many of the main features. Within the scheme, points are awarded for each biological, aesthetic, and water quality feature within each part of the estuary and then the points are totalled for each part. The points are then related to a Class indicating good quality (Class A), fair quality (Class B), poor quality (Class C) to bad quality (Class D).

Following the classification of an area, where the area within each class can be summarized for an estuary, targets can then be set for future improvements or remediation. Similarly, if the scheme is used at repeated intervals, say 5 years, then it should indicate the effectiveness or even cost–benefit analysis of protection measures employed. For example, in the Forth estuary, Scotland, an area around the Grangemouth oil refinery was originally designated as Class C because of the changes to the benthos resulting from the refinery effluent discharged to the intertidal area. The building of a wastewater treatment plant and extension to the effluent pipe to increase the dilution of the waste then resulted in the quality of the area being improved to Class B such that the refinery company and the local EPA were able to quantify in area the benefits of spending money on the new plant.

As a wider form of classification systems and geographical assessments, there is an increasing use of Quality Status Reports (QSR)

Box 8.7 Estuarine Classification Scheme (Produced by the Environment and Heritage Service, Northern Ireland)

Biological Quality: points awarded for the passage of migratory fish, the support of a usual and expected fish community, the presence of a typical benthic community, the ability to meet microbiological standards, and the absence of toxic/tainting substances in biota.
Aesthetic Quality: points based on the easily observed signs of estuarine polluting inputs sufficient to interfere with estuarine usage or to cause aesthetic pollution; includes the presence on beaches and in the waters of litter, sewage solids, foaming, slicks, etc.
Water quality: points based on the ability to meet agreed (nationally or at EU level) chemical standards such as ammonia or dissolved oxygen concentrations exceeding specified saturation values, levels of persistent pollutants.
Summation of points in each category to produce the Estuary Classification Scheme:

Class	Description	Aesthetic condition	Fish migration	Resident biota and/or bioassay	Resident fish	Persistent substances (Biota)	Dissolved oxygen (DO) in water	UK Red list and EC dangerous substances in water
A	Excellent	Unpolluted	Water quality allows free passage	Normal	Resident fish community normal	<Guideline Standard NI Shellfish Waters	Minimum DO > 6 mg l^{-1}	100% compliance of samples with EQS
B	Good	May show signs of contamination	Water quality allows free passage	Normal	Resident fish community normal	< Mandatory but > Guideline Standard NI Shellfish Waters	Minimum DO < or = 6 mg l^{-1} but >4 mg l^{-1}	Annual compliance of samples with EQS
C	Unsatisfactory	Occasional observations or substantiated complaints of pollution	Water quality restricts passage	Modified	Resident fish community modified	>Mandatory Standard NI Shellfish Waters	Minimum DO < or = 4 mg l^{-1} but >2 mg l^{-1}	One or more List II substances fail to comply with EQS. List I and Red List all comply
D	Seriously polluted	Frequent observations or substantiated complaints of pollution	Water quality prevents passage	Impoverished or seriously modified	Resident fish community impoverished	> 2X Mandatory Standard NI Shellfish Waters	DO <2 mg l^{-1}	One or more List I or Red List substances fail to comply with EQS

A given area of estuary is classified by allocating it to the highest class to which all of its criteria conform. An estuarine are satisfying Class B aesthetic and chemical criteria but which is Class C on the basis of one of the biological criteria would be classified as Class C overall (i.e. a default system). If there are only limited biological and chemical data then estuaries will be classified using information for known discharges, pollution complaints, etc. If there are none of the latter then the area will be Class A.

such as those produced by the OSPAR for the North Sea and the NE Atlantic area. These QSR summarize the nature of these sea areas, their natural and anthropogenic features, and the threats to their resources.

8.5 Final remarks

Although the discussion here has indicated methods for dealing with pollution, and especially that from point sources such as dumping from vessels or from pipelines, the more intractable problems requiring management of estuaries are diffuse pollution, such as from farmland, and the loss of habitats. The problems of the destruction of estuarine habitat by reclamation and other engineering works may, in the long run, prove to be the most intractable in managing our estuarine ecosystems. Management has to include an assessment of the impact of loss of area, not only as a food loss for birds or fish, but also for the loss of a "free sewage-works" area, which can cope with the various pollutants discussed above. It is essential to ask appropriate question and to understand the functioning of all aspects of the area before undertaking any irreversible activities (Fig. 8.5).

The conservation interest in estuaries is focused on two aspects. First the maintenance of fisheries, either by the harvesting of fish and shellfish, or by the use of estuaries as

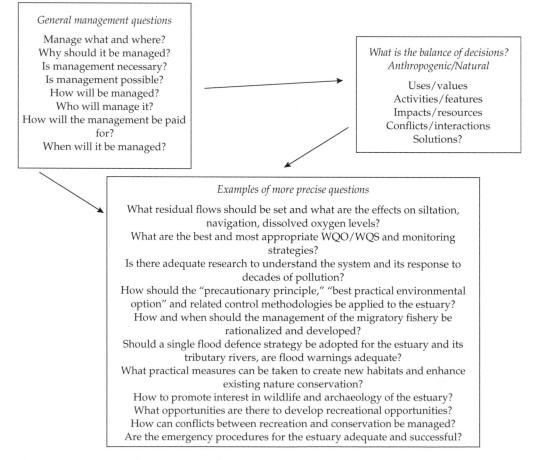

General management questions

Manage what and where?
Why should it be managed?
Is management necessary?
Is management possible?
How will be managed?
Who will manage it?
How will the management be paid for?
When will it be managed?

What is the balance of decisions?
Anthropogenic/Natural

Uses/values
Activities/features
Impacts/resources
Conflicts/interactions
Solutions?

Examples of more precise questions

What residual flows should be set and what are the effects on siltation, navigation, dissolved oxygen levels?
What are the best and most appropriate WQO/WQS and monitoring strategies?
Is there adequate research to understand the system and its response to decades of pollution?
How should the "precautionary principle," "best practical environmental option" and related control methodologies be applied to the estuary?
How and when should the management of the migratory fishery be rationalized and developed?
Should a single flood defence strategy be adopted for the estuary and its tributary rivers, are flood warnings adequate?
What practical measures can be taken to create new habitats and enhance existing nature conservation?
How to promote interest in wildlife and archaeology of the estuary?
What opportunities are there to develop recreational opportunities?
How can conflicts between recreation and conservation be managed?
Are the emergency procedures for the estuary adequate and successful?

Figure 8.5 Questions in the management of an estuary.

nursery areas for marine fish stocks, or for fish such as sprat and flounder caught within the estuary, or for estuaries as routes for migratory species to pass through. The conservation of fish populations is inhibited by poor water quality (usually low dissolved oxygen, but also, e.g., high ammonia levels), which acts as a barrier to migratory fish, as well as levels of persistent pollutants, and high nutrient levels, which after the collapse of a phytoplankton bloom can lead to low dissolved oxygen levels. Second, the conservation interest is centered on the abundant bird populations, which feed on the intertidal areas, attracted by the rich and available food supplies. To protect these interests, many estuaries have been declared as nature reserves, primarily because, as stated in the original paradoxes, they can be some of the least disturbed natural areas with rich natural animal populations, and also to help protect them against pollution or reclamation, which can destroy them forever. The conservation of wildlife in estuaries need not exclude other users. Industries are compatible with wildlife, provided that organic effluents and power-station discharges are well diluted and carefully sited. Indeed small amounts of organic effluents, and well-controlled power-station discharges may actually serve to increase the productivity of an estuary, and enhance its conservation interest. Persistent chemicals, which accumulate in the food chain must, however, be strictly controlled. Excessive reclamation, especially if it takes away intertidal areas, which are often more productive than subtidal areas, poses probably the greatest single threat to estuarine conservation, destroying forever the habitat.

The management of estuaries is thus largely the protection of biological systems, even if the biology is that of mankind. The future management of estuaries as a resource for wildlife as well as for mankind depends very much on the successful implementation of legislation for pollution control and habitat protection. As emphasized throughout the chapters here, the options for sustainable and cost-effective management have to fulfil six tenets: being environmentally sustainable, economically viable, technologically feasible, legally permissible, socially desirable, and administratively achievable. Because of this, an estuarine scientist working on real environmental problems cannot ignore all the aspects mentioned here; similarly, managers, legislators, and policy-makers need an understanding of the science of estuaries in order to interpret and implement the available tools. Only by taking such a wide view can our estuaries be managed properly for the benefit of the wildlife and Man.

Reading list

The reading list for this book consists of first a selection of general reference books, multi-author works and journals which have applicability to all of the book, or to several chapters, followed by references to individual papers which are arranged in chapter groupings. If a reference relates to more than one chapter it is listed in the first chapter in which appears.

Books and Multi-author works

Allanson, B. R. and Baird, D. (eds.) (1999). *Estuaries of South Africa*. Cambridge University Press, Cambridge, 340 pp.

Aller, J. Y., Woodin, S. A., and Aller, R. C. (eds.) (2001). *Organism-sediment Interactions*. University of South Carolina Press, Columbia, S.C. Belle W Baruch Library in marine science, Number 21, 403 pp.

Attrill, M. J. (ed.) (1998). *A Rehabilitated Estuarine Ecosystem*. Kluwer, Dordrecht. 254 pp.

Baretta-Bekker, J. G., Duursma, E. K. and Kuipers, B. R. (eds.) (1998). Encyclopaedia of Marine Sciences. *(second edition)* Springer-Verlag, Berlin, 357 pp.

Barnes, M. and Gibson, R. N. (eds.) (1990). *Trophic Relationships in the Marine Environment*. Aberdeen University Press, Aberdeen, 642 pp.

Barnes, R. S. K. (1994). *The brackish-water Fauna of NW Europe*. Cambridge University Press, Cambridge, 287 pp.

Barnes, R. S. K. and Hughes, R. N. (1999). *An Introduction to Marine Ecology, (third edition)* Blackwells, Oxford, 286 pp.

Bayne, B. L., Clarke, K. R. and Gray, J. S. (eds.) (1988). Biological Effects of Pollutants. *Marine Ecology Progress Series* Vol. **46** (1–3), 1–278.

Bell, S. and McGillivray, D. (2000). *Environmental Law (fifth edition of Ball and Bell)*. Blackstone Press, London, 726 pp.

Bergman, H. L., Kimerle, R. A. and Maki, A.W. (eds.) (1986). *Environmental Hazard Assessments of Effluents*. Pergamon Press, Oxford and New York, 366 pp.

Bird, E. C. F. (1993). *Submerging coasts – The Effects of a Rising Sea Level on Coastal Environments*. John Wiley and Sons, Chichester, 194 pp.

Bird, E. C. F. (1996). *Beach Management*. John Wiley and Sons, Chichester, 281 pp.

Bird, E. C. F. (2000). *Coastal Geomorphology: An Introduction*. John Wiley and Sons, Chichester, 322 pp.

Birnie, P. W. and Boyle A. E. (1992). *International Law and the Environment*, Oxford University Press, Oxford, 563 pp.

Carter, R. W. G. (1988). *Coastal Environments – An Introduction to Physical, Ecological and Cultural Systems of Coastlines*. Academic Press, London, 640 pp.

Chapman, D. (ed.) (1996). *Water Quality Assessment: A Guide to the Use of Biota, Sediments and Water in Environmental Monitoring. (2nd edition)* E. and F. N. Spon, London, 626 pp.

Chapman, V. J. (ed.) (1977). *Wet Coastal Ecosystems*. Elsevier, Amsterdam, 428 pp.

Churchill, R. R. and Lowe, A. V. (1999). *The Law of the Sea. (3rd edition)* Manchester University Press, Manchester, 494 pp.

Cicin-Sain, B. and Knecht, R. W. (1998). *Integrated Coastal and Ocean Management: Concepts and Practices*. Island Press, Washington DC, 517 pp.

Clark, R. B. *et al.* (2001). *Marine Pollution (fifth edition)*. Oxford University Press, Oxford, 237 pp.

Constable, A. J. and Fairweather, P. G. (eds.) (1999). Ecology of estuaries and soft-sediment habitats. *Australian Journal of Ecology*, **24**, 289–476.

Davidson, N. C. *et al.* (1991). *Nature Conservation and Estuaries in Great Britain*. Nature Conservancy Council, Peterborough, 422 pp.

Day, J. H. (ed.) (1980). *Estuarine ecology, with particular reference to southern Africa*. Balkema books, Rotterdam, 400 pp.

Day, J. W. *et al.* (1989). *Estuarine Ecology.* John Wiley, New York, 558 pp.

De Leeuw, C. C. and Backx, J. J. G. M. (2001). *Naar een hertsel van estuariene gradienten in Nederland.* Rijkswaterstaat rapport nr. 2000.044, 167 pp.

Department of the Environment (1993). *Coastal Planning and Management: A Review.* HMSO Publications, London, 178 pp.

Dickson, K. R., Maki, A. W. and Brungs, W. A. (eds.) (1987). *Fate and Effects of Sediment-bound Chemicals in Aquatic Systems.* Pergamon Press, New York, 449 pp.

Dyer, K. R. (1986). *Coastal and Estuarine Sediment Dynamics.* Wiley, Chichester, 342 pp.

Dyer, K. R. (1997). *Estuaries – A Physical Introduction (2nd Edition).* John Wiley, Chichester, 199 pp.

Dyer, K. R and Orth, R. J. (eds.) (1994). Changes in fluxes in estuaries: implications from science to management. Proceedings of the ECSA22/ERF Symposium Olsen and Olsen, Fredensborg, Denmark. 485 pp.

Dyke, P. (1996). *Modelling Marine Processes.* Prentice-Hall, London, 152 pp.

Elliott, M. and Ducrotoy, J-P. (eds.) (1991). *Estuaries and Coasts: Spatial and Temporal Intercomparisons.* Olsen and Olsen, Fredensborg, Denmark. 390 pp.

Elliott, M. and Ducrotoy J P. (eds) (1994). Environmental Perspectives for the Northern Seas. *Marine Pollution Bulletin*, **29**, 253–660.

French, P. (1997). *Coastal and Estuarine Management.* Routledge, London, 251 pp.

Goksøyr, A. (ed.) (1998). Pollution Responses in Marine Organisms (PRIMO9). *Marine Environmental Research*, Special Issue, **46**, 1–607.

Harrop, D. O. and Nixon, J. A. (1999). *Environmental Assessment in Practice.* Routledge, London, 219 pp.

Heip, C. H. R. *et al.* (1995). Production and consumption of biological particles in temperate tidal estuaries. *Oceanography and Marine Biology Annual Review*, **33**, 1–149.

Heip, C. H. R. and Herman, P. M. J. (eds.) (1995). Major Biological Processes in European Tidal estuaries. *Hydrobiologia*, **311**, 1–266.

Hobbie, J. E. (ed.) (2000). *Estuarine Science: a synthetic approach to research and practices.* Island Press, Washington, 539 pp.

Jones, M. B. and Millward, G. E. (eds.) (2002). Strategic estuarine research in the new millennium. *Estuarine Coastal and Shelf Science*, **55**, 805–1008.

Jones, N. V. and Wolff, W. J. (eds.) (1981). *Feeding and Survival Strategies in Estuarine Organisms.* Plenum Press, New York. 304 pp.

Kay, R. and Alder, J. (1999). *Coastal Planning and Management.* E. and F. N. Spon (Routledge), London, 375 pp.

Kennedy, V. S. (ed.) (1980). *Estuarine Perspectives.* Academic Press, New York, 533 pp.

Kennedy, V. S. (ed.) (1982). *Estuarine Comparisons.* Academic Press, New York, 709 pp.

Kennedy, V. S. (ed.) (1984). *The Estuary as a Filter.* Academic Press, New York.

Kennish, J. R. (1986). *Ecology of Estuaries (Volume 1, Physical and Chemical aspects).* CRC Press, Florida, 254 pp.

Kennish, M. J. (ed.) (2000). *Estuary Restoration and Maintenance.* CRC Press, Florida.

Ketchum, B. H. (ed.) (1983). *Estuaries and Enclosed Seas.* Elsevier, Amsterdam, 500 pp.

Kinne, O. (ed.) (1978, onwards). *Marine Ecology* (5 volumes) John Wiley, New York.

Kjerfve, B. (ed.) (1978). *Estuarine Transport Processes.* Belle W. Baruch Library in Marine Science, No. 7, University of South Carolina Press, Columbia, S.C., 331 pp.

Knox, G. A. (1986). *Estuarine Ecosystems (2 volumes).* CRC Press, Florida, 289 + 230 pp.

Lalli, C. M. and Parsons, T. R. (1997). *Biological Oceanography: An Introduction. (second edition)* Butterworth Heinemann, Oxford, 314 pp.

Lauff, G. H. (ed.) (1967). *Estuaries.* American Association for the advancement of science, Washington, D.C. Publication number 83, 755 pp.

Lawrence, A. J. and Hemingway, K. L. (eds.) (2003). *Effects of Pollution on Fish.* Blackwell Science Publications, Oxford, 342 pp.

Libes, S. M. (1992). *An Introduction to Marine Biogeochemistry.* John Wiley, New York, 734 pp.

Little, C. (2000). *The Biology of Soft Shores and Estuaries.* Oxford University Press, Oxford, 252 pp.

Mann, K. H. (1982). *Ecology of Coastal Waters: A Systems Approach.* Blackwell Science Publications, Oxford, 322 pp.

McDowall, R. M. (1988). *Diadromy in Fishes.* Croom-Helm, London, 308 pp.

McManus, J. and Elliott, M. (eds.) (1989). *Developments in Estuarine and Coastal Study Techniques.* Olsen and Olsen, Fredensborg, 158 pp.

McLusky, D. S. (ed.) (1987). Symposium on the Natural Environment of the Estuary and Firth of

Forth. *Proceedings of the Royal Society of Edinburgh*, **93B**, 235–571.

McLusky, D. S., de Jonge, V. N. and Pomfret, J. (eds.) (1990). *North Sea – Estuaries Interactions. Hydrobiologia*, **195**, 1–221.

Meire, P and Vincx, M. (eds.) (1993). Marine and estuarine gradients. *Netherlands Journal of Aquatic Ecology*, **27**, 71–496.

Moreira, M. H., Quintino, V. M. S. and Rodrigues, A. M. J. (eds.) (1995). Northern and southern estuaries and coastal areas. *Netherlands Journal of Aquatic Ecology*, **29**, 199–471.

Moulder, D. S. and Williamson, P. (eds.) (1986). Estuarine and Coastal Pollution: Detection, Research and Control. *Water Science and Technology*, **18**, 1–364.

National Research Council (2000). *Clean Coastal Waters – Understanding and Reducing the Effects of Nutrient Pollution*. National Academy Press, Washington, D.C., 405 pp.

Nedwell, B. D. and Raffaelli, D. G. (eds.) (1999). Estuaries. *Advances in Ecological Research*, **29**. 1–295, Academic Press, London.

Neff, J. M. (2002). *Bioaccumulation in Marine Organisms – Effect of Contaminants from Oilwell Produced Water*. Elsevier, Amsterdam, 452 pp.

Newman, P. J. and Agg, A. R. (eds.) (1988). *Environmental Protection of the North Sea*. Heinemann, Oxford, 886 pp.

Nienhuis, P. H. and Smaal, A. C. (eds.) (1994). *The Oosterschelde Estuary (The Netherlands): A Case Study of a Changing Ecosystem*. Kluwer Academic Publishers, Dordrecht, 597 pp.

Open University Course Team (1999). *Waves, Tides and Shallow-water Processes. (second edition)*. Open University, Butterworth-Heinemann, Oxford, 227 pp.

Orive, E., Elliott, M. and de Jonge, V. N. (eds.) (2002). *Nutrients and Eutrophication in Estuaries and Coastal Waters*. Kluwer Academic Publishers, Dordrecht, 526 pp.

Penning-Rowsell, E. *et al.* (1992). *The Economics of Coastal Management*, Belhaven Press, London, 380 pp.

Perkins, E. J. (1974), *The Biology of Estuarine and Coastal Waters*. Academic Press, London, 678 pp.

Pethick, J. S. (1984). *An introduction to coastal geomorphology*. Arnold, London, 260 pp.

Phillips, D. J. H. and Rainbow, P. S. (1994). *Biomonitoring of Trace Aquatic Contaminants*. Chapman and Hall, London, 371 pp.

Postma, H. and Zijlstra, J. J. (eds.) (1988). *Continental shelves*. Elsevier, Amsterdam, 421 pp.

Raffaelli, D. and Hawkins, S. (1996). *Intertidal Ecology*. Chapman and Hall, London, 356 pp.

Raymont, J. E. G. (1980). *Plankton and Productivity in the Oceans (Volume 1: Phytoplankton). (second edition)*. Pergamon Press, Oxford, 489 pp.

Raymont, J. E. G. (1983). *Plankton and Productivity in the Oceans (Volume 2: Zooplankton). (second edition)*. Pergamon Press, Oxford, 824 pp.

Reise, K. (1983). *Tidal Flat Ecology – An Experimental Approach*. Springer-Verlag, Heidelberg, 191 pp.

Reise, K. (2001). *Ecological Comparisons of Sedimentary shores*. Springer, Heidelberg, 387 pp.

Rijkswaterstaat (1985). *Biological Research in the Ems-dollart Estuary*. Rijkswaterstaat, The Hague, Netherlands, 182 pp.

Ros, J. D. (ed.) (1989). *Topics in Marine Biology (Proceedings of 23rd EMBS). Scientia Marina*, **53 (2–3)**, 145–754.

Salomons, W., Bayne, B. I. Duursma, E. K. and Forstner, U. (eds.) (1988). *Pollution of the North Sea – An Assessment*. Springer-Verlag, Berlin, 687 pp.

Schmitt, R. J. and Osenberg, C. W. (1996). *Detecting Ecological Impacts: Concepts and Applications in Coastal Habitats*. Academic Press, New York, 401 pp.

Schramm, W. and Nienhuis, P. H. (eds.) (1996). *Marine Benthic Vegetation: Recent Changes and the Effects of Eutrophication*. Springer-Verlag, Berlin, 470 pp.

Sheppard, C. (ed.) (2000). *Seas at the Millennium: An Environmental Evaluation*. 3 volumes, Pergamon Press, Oxford, 2400 pp.

Steele, J. H., Turekian, K. K. and Thorpe, S. A. (eds.) (2001). *Encyclopaedia of Ocean Sciences*, 6 Volumes. Academic Press, London, 3399 pp.

Tait, R.V. and Dipper, F. A. (1998). *Elements of Marine Ecology (4th edition)*. Butterworth-Heinemann, Oxford, 462 pp.

Tenore, K. R. and Coull, B. C. (eds.) (1980). *Marine Benthic Dynamics*. Belle W. Baruch Library in Marine Science, No. 11, University of South Carolina Press, Columbia, S.C., 451 pp.

Viles, H. and Spencer, T. (1995). *Coastal Problems: Geomorphology, Ecology and Society at the Coast*. Edward Arnold, London, 350 pp.

Walker, C. H., Hopkin, S. P., Sibly R. M. and Peakall D. B. (1996). *Principles of Ecotoxicology*. Taylor and Francis, London, 321 pp.

Whitehouse R., Soulby, R., Roberts, R. and Mitchener, H. (2000). *Dynamics of Estuarine Muds.* Thames Telford Ltd., London, 210 pp.

Wiley, M. (ed.) (1976). *Estuarine Processes* 2 volumes. Academic Press, London, 603 pp.

Wilson, J. G. and Halcrow, W. (eds.) (1985). *Estuarine Management and Quality Assessment.* Plenum Press, New York, 225 pp.

Wilson, J. G. (1983). *The Biology of Estuarine Management.* Croom Helm, London. 204 pp.

Wolfe, D. A. (ed.) (1986). *Estuarine Variability.* Academic Press, Florida, 509 pp.

Wolff, W. J. (ed.) (1980). *Ecology of the Wadden Sea,* 3 volumes. Balkema Books, Rotterdam, 1300 pp.

Handbooks on methods for studying estuaries

Bagenal, T. B. (1978). *Methods for Assessment of Fish Production in Fresh Waters.* Blackwell Scientific Publications, Oxford. 3rd edition, 365 pp.

Baker, J. M. and Wolff, W. J. (eds.) (1987). *Biological Surveys of Estuaries and Coasts.* EBSA handbook, Cambridge University Press, Cambridge, 449 pp.

Dyer, K. R. (ed.) (1979). *Estuarine Hydrography and Sedimentation.* EBSA handbook, Cambridge University Press, Cambridge, 230 pp.

Head, P. C. (ed.) (1985). *Practical Estuarine Chemistry.* EBSA handbook, Cambridge University Press, Cambridge, 337 pp.

Holme, N. A. and McIntyre, A. D. (1984). *Methods for the Study of Marine Benthos (2nd edition).* Blackwell Scientific Publications, Oxford, 387 pp.

Kramer, K. J., Brockmann, U. H. and Warwick, R. M. (1994). *Tidal Estuaries: Manual of Sampling and Analytical Procedures.* AA Balkema, Rotterdam, 304 pp.

Morris, A. W. (ed.) (1983). *Practical Procedures for Estuarine Studies.* Natural Environment Research Council, Plymouth, 262 pp.

Journals

The publication of material relating to the Estuarine Ecosystem has been greatly stimulated by the Estuarine and Coastal Sciences Association (ECSA, mainly for United Kingdom and Europe) and the Estuarine Research Federation (ERF, mainly for North America) who organize regular meetings and publish bulletins, symposium volumes and other publications. Any reader who would like details of these associations is invited to contact the authors who will forward the current address of the ECSA or ERF Secretary.

These organizations produce or sponsor the key journals in this subject area: *Estuarine Coastal and Shelf Science* and *Estuaries.*

In addition to the above journals, the following journals contain a selection of relevant papers.

Aquatic Ecology
Aquatic Science and Biology Abstracts
Coastal Zone Topics
Journal of Animal Ecology
Journal of Applied Ecology
Journal of Experimental Marine Biology and Ecology
Journal of the Marine Biological Association of the U.K.
Limnology and Oceanography
Marine Biology
Marine Ecology Progress Series
Marine Pollution Bulletin
Journal of Sea Research
Oceanography and Marine Biology. An Annual Review
Ophelia

Chapter One

Anderson, F. E. (1983). The northern muddy intertidal: a seasonally changing source of suspended sediments to estuarine waters-a review. *Canadian Journal of Fisheries and Aquatic Sciences,* **40** (Supp. 1), 143–59.

Barnes, R. S. K. (1980). *Coastal Lagoons.* Cambridge University Press, Cambridge. 106 pp.

Bale, A. J. and Morris, A. W. (1987). *In situ* measurement of particle size in estuarine waters. *Estuarine, Coastal and Shelf Science,* **24**, 253–63.

Biggs, R. B. (1970). Sources and distribution of suspended sediment in North Chesapeake Bay. *Marine Geology,* **9**, 187–201.

Burton, J. D. and Liss, P. S. (eds.) (1976). *Estuarine chemistry.* Academic Press, London, 229 pp.

Elliott, M. and McLusky, D. S. (2002). The need for definitions in understanding estuaries. *Estuarine Coastal and Shelf Science* **55**, 815–27.

El-Sabh, M. I., Augn, T. H. and Murty, T. S. (1997). Physical processes in inverse estuarine systems. *Oceanography and Marine Biology: An annual review,* **35**, 1–69.

Essink, K. and de Jonge, V. N. (eds.) (1994). Particles in estuaries and coastal waters. *Netherlands Journal of Aquatic Ecology,* **28**, 233–478.

Fairbridge, R. (1980). The estuary: its definition and geodynamic cycle. In E. Olausson and I. Cato (eds.), *Chemistry and Geochemistry of Estuaries*, 1–35. John Wiley and Son, New York.

Gray, J. S. (1981). *The Ecology of Marine Sediments*. Cambridge University Press, Cambridge, 185 pp.

Head, P. C. (1976). Organic processes in estuaries. In J. D. Burton and P. S. Liss, (eds.) *Estuarine Chemistry*. Academic Press, London, 54–91.

Hedgpeth, J. W. (1967). The sense of the meeting. In G. H. Lauff (ed.) *Estuaries*, American Association for the Advancement of Science, Washington, D.C., **83**, 707–12.

King, C. M. (1975). *Introduction to Marine Geology and Geomorphology*. E. Arnold, London, 370 pp.

Krone, R. B. (1978). Aggregation of suspended particles in estuaries. In B. Kjerfve, (ed.) *Estuarine Transport Processes*. University of South Carolina Press, Columbia, S.C., 177–90.

Jarvis, J. and Riley, C. (1987). Sediment transport in the mouth of the Eden estuary *Estuarine, Coastal and Shelf Science*, **24**, 463–81.

Mantoura, R. F. C. (1987). Organic films at the halocline. *Nature*, **328**, 579–80.

McLusky, D. S. (1993). Marine and estuarine gradients – an overview. *Netherlands Journal of Aquatic Ecology*, **27**, 489–93.

Middelburg, J. J. and Nieuwenhuize, J. (2000). Uptake of dissolved inorganic nitrogen in turbid, tidal estuaries. *Marine Ecology Progress Series*, **192**, 79–88.

Middelburg, J. J. and Nieuwenhuize, J. (2001). Nitrogen isotope tracing of dissolved inorganic nitrogen behaviour in tidal estuaries. *Estuarine Coastal and Shelf Sciences*, **53**, 385–91.

Morris, A. W. *et al.* (1978). Very low salinity regions of estuaries: important sites for chemical and biological reactions *Nature*, **274**, 678–80.

Nienhuis, P. H. (1992). Eutrophication, water management, and the functioning of Dutch estuaries and coastal lagoons. *Estuaries*, **15**, 538–48.

Nienhuis, P. H. (1993). Nutrient cycling and food-webs in Dutch estuaries. *Hydrobiologia*, **265**, 15–44.

Paterson, D. M. and Black, K. S. (1999). Water flow, sediment dynamics and benthic biology. *Advances in Ecological research*, **29**, 155–94.

Pearson, T. and Stanley, S. O. (1979). Comparative measurement of the redox potential of marine sediments as a rapid means of assessing the effect of organic pollution. *Marine Biology*, **53**, 371–79.

Phillips, J. (1972). Chemical processes in estuaries. In R. S. K.Barnes and J. Green, (eds.), *The Estuarine Environment*. Applied Science Publishers, London, pp.33–50.

Postma, H. (1967). Sediment transport and sedimentation in the estuarine environment. In G.H.Lauff, (ed.), *Estuaries*. American Association for the Advancement of Science, Washington, D.C., Publication No. 83, 158–79.

Pritchard, D. W. (1967). What is an estuary: a physical viewpoint. In G.H.Lauff, (ed.), *Estuaries*. American Association for the Advancement of Science, Washington, D.C., Publication No. 83, 3–5.

Rhoads, D. C. (1974). Organism-sediment relations on the muddy sea floor. *Oceanography and Marine Biology, Annual Review*, **12**, 263–300.

Roy, P.S. *et al.* (2001). Structure and function of southeast Australian estuaries. *Estuarine Coastal and Shelf Science*, **53**, 351–84.

Sholkovitz, E. R. (1979). Chemical and physical processes controlling the chemical composition of suspended material in the River Tay estuary. *Estuarine and Coastal Marine Science*, **8**, 523–45.

Soetaert, K. and Herman, P. M. J. (1995a). Nitrogen dynamics in the Westerschelde estuary (SW Netherlands) estimated by means of the ecosystem model MOSES. *Hydrobiologia*, **311**, 225–46.

Soetaert, K. and Herman, P. M. J. (1995b). Carbon flows in the Westerschelde estuary (SW Netherlands) estimated by means of the ecosystem model MOSES. *Hydrobiologia*, **311**, 247–66.

Romao, C. (1996). *Interpretation Manual of European Union Habitats*, Version EUR15. European Commission, DGXI (Environment, Nuclear Security and Civil Protection), Brussels, 106 pp.

Uncles, R. J. (2002). Estuarine physical processes research: some recent studies and progress. *Estuarine Coastal and Shelf Science*, **55**, 829–56.

Tappin, A. D. (2002). An examination of the fluxes of nitrogen and phosphorus in temperate and tropical estuaries: current estimates and uncertainties. *Estuarine Coastal and Shelf Science*, **55**, 885–906.

Turner, A. and Millward, G. E. (2002). Suspended particles: their role in estuarine Biogeochemical cycles. *Estuarine Coastal and Shelf Science*, **55**, 857–84.

Van Beusekom, J. E. E. and de Jonge, V. N. (1998). Retention of phosphorus and nitrogen in the Ems estuary. *Estuaries*, **21**, 527–39.

Chapter Two

Alexander, W. B., Southgate, B. A. and Bassindale, R. (1935). Survey of the River Tees. Part II—the estuary, chemical and biological. *D.S.I.R., Water Pollution Research Technical Paper*, **5**, 1–171.

Beukema, J. (1997). Caloric values of marine invertebrates with an emphasis on the soft parts of marine bivalves. *Oceanography and Marine Biology: An Annual Review* **35**, 387–412.

Brey, T., Rumohr, H. and Ankar, S. (1988). Energy content of macrobenthic invertebrates: general conversion factors from weight to energy. *Journal of Experimental Marine Biology and Ecology*, **117**, 271–8.

Buchanan, J. B. and Warwick, R. M. (1974). An estimate of benthic macrofaunal production in the offshore mud of the Northumberland Coast. *Journal of the Marine Biological Association of the UK*, **54**, 197–222.

Carriker, M. R. (1967). Ecology of estuarine benthic invertebrates: a perspective. In G.H. Lauff (ed.) *Estuaries.* American Association for the Advancement of Science, Washington, D.C. Publication No. 83, 442–87.

Chambers, M. R. and Milne, H. (1979). Seasonal variation in the condition of some intertidal invertebrates of the Ythan estuary *Estuarine and Coastal Marine Science*, **8**, 411–19.

Crisp, D. (1984). Energy flow measurements. In Holme, N.A. and McIntyre, A.D. (eds.) *Methods for Studying the Marine Benthos* 2nd edition. IBP handbook 16, Blackwell, Oxford, 284–372.

Deaton, L. E. and Greenberg, M. J. (1986). There is no horohalinicum. *Estuaries*, **9**, 20–30.

Hughes, R. N. (1970). An energy budget for a tidal flat population of the bivalve *Scrobicularia plana. Journal of Animal Ecology*, **39**, 357–81.

Khayralla, N. and Jones, A. M. (1975). A survey of the benthos of the Tay estuary. *Proceedings of the Royal Society of Edinburgh*, **75B**, 113–35.

Khlebovich, V. V. (1968). Some peculiar features of the hydrochemical regime and the fauna of mesohaline waters. *Marine Biology*, **2**, 47–9.

Knox, G. A. and Kilner, A. R. (1973). *The Ecology of the Avon-Heathcote Estuary.* University of Canterbury, New Zealand, 358 pp.

Lindemann, R. L. (1942). The trophic-dynamic aspect of ecology. *Ecology*, **23**, 399–418.

Lippson, A. J. *et al.* (1981). *Environmental Atlas of the Potomac Estuary.* John Hopkins University Press, Maryland. 280 pp.

Maitland, P. S. and Hudspith, P. M. G. (1984). The zoobenthos of Loch Leven, Kinross and estimates of its production in the sandy littoral area during 1970–71. *Proceedings of the Royal Society of Edinburgh*, **74B**, 219–40.

McIntyre, A. D. (1970). The range of biomass in intertidal sand, with special reference to *Tellina tenuis. Journal of the Marine Biological Association of the UK*, **50**, 561–76.

McNeill, S. and Lawton, J. H. (1970). Annual production and respiration in animal populations. *Nature*, **225**, 472–4.

Muus, B. (1974). Ecophysiological problems of the brackish water. *Hydrobiological Bulletin*, **8**, 76–89.

Rankin, J. C. and Davenport, J. A. (1981). *Animal Osmoregulation*. Blackie, Glasgow, 220 pp.

Remane, A. and Schlieper, C. (1958). *Die biologie des brackwassers.* E. Schwiezerbart'sche verlagsbuchhandlung, Stuttgart, 348 pp.

Salonen, K., Sarvala, J., Hakala, I. and Viljanen, M.-J. (1976). The relation of energy and organic carbon in aquatic invertebrates. *Limnology and Oceanography*, **21**, 724–30.

Sanders, H. L. (1969). Benthic marine diversity and the stability-time hypothesis. *Brook-haven Symposia in Biology*, **22**, 71–81.

Schwinghamer, P. *et al.* (1986). Partitioning of production and respiration among size groups of organisms in an intertidal benthic community. *Marine Ecology Progress Series*, **31**, 131–42.

Venice System (1959). Symposium on the classification of brackish waters, Venice, April 8–14, 1958. *Arch. Oceanog. Limnol.*, **11** (supplement), 1–248.

Waters, T. F. (1969). The turnover ratio in production ecology of freshwater invertebrates. *American Naturalist*, **103**, 173–85.

Wildish, D. J. (1977). Factors controlling marine and estuarine sublittoral macrofauna. *Helgolander wiss. Meeresunters.*, **30**, 445–54.

Wilson, J.G. (2002). Productivity, Fisheries and aquaculture in Temperate Estuaries. *Estuarine Coastal and Shelf Science*, **55**, 953–68.

Winberg, G. C. (1971). *Methods for the Estimation of the Production of Aquatic Animals*. Academic Press, London, 175 pp.

Wolff, W. J. (1972). Origin and history of the brackish-water fauna of NW. Europe. *Proc. 5th Europe. Mar. Biol. Symp*, Piccin Ed., Padova, 11–18.

Chapter Three

Andersen, T. J. (2001). Seasonal variation in the erodibility of two temperate microtidal mudflats. *Estuarine, Coastal and Shelf Sciences*, **53**, 1–12.

Baillie, P. W. (1986). Oxygenation of intertidal estuarine sediments by benthic microalgal photosynthesis. *Estuarine, Coastal and Shelf Science*, **22**, 143–59.

Black, K. S., Patterson, D. M. and Cramp, A. (eds.) (1998). *Sedimentary Processes in the Intertidal Zone*. Geological Society, London, Special Publications, 139, 409 pp.

Boynton, W. R., Kemp, W. M. and Keefe, C. W. (1982). A comparative analysis of nutrients and other factors influencing estuarine phytoplankton production. In V. S. Kennedy, (ed.) *Estuarine Comparisons*. Academic Press, New York, 69–90.

Cadee, G. C. (1971). Primary production on a tidal flat. *Netherlands Journal of Zoology*, **21**, 213–25.

Cattrijsse, A., Dankwa, H. R. and Mees, J. (1997). Nursery function of an estuarine tidal marsh for the brown shrimp, *Crangon crangon*. *Journal of Sea Research*, **38**, 109–21.

Cloern, J. E. (1982). Does the benthos control phytoplankton biomass in south San Francisco Bay? *Marine Ecology, Progress Series*, **9**, 191–202.

Colijn, F. and de Jonge, V. N. (1984). Primary production of microphytobenthos in the Ems-Dollard estuary. *Marine Ecology, Progress Series*, **14**, 185–96.

Dame, R. F. and Stilwell, D. (1984). Environmental factors influencing macrodetritus flux in North Inlet estuary. *Estuarine, Coastal and Shelf Science*, **18**, 721–6.

Darnell, R. M. (1967). The organic detritus problem. In G. H. Lauff, (ed.) *Estuaries*. American Association for the Advancement of Science, Washington, D.C., Publication No. 83, 374–5.

Dame, R. F. and Allen, D. M. (1996). Between estuaries and the sea. *Journal of Experimental Marine Biology and Ecology*, **200**, 169–85.

Duke, N. C. (1992). Mangrove floristics and biogeography. In A. I. Robertson and D. Alongi (eds.), *Tropical Mangrove Ecosystems*. American Geophysical Union, Washington, D.C., 63–100.

Eilers, H. P. (1979). Production ecology in an Oregon coastal salt marsh. *Estuarine and Coastal Marine Science*, **8**, 399–410.

Gallacher, J. L., Pfeiffer, W. J. and Pomeroy, L. R. (1976). Leaching and microbial utilisation of dissolved organic carbon from leaves of *Spartina alterniflora*. *Estuarine and Coastal Marine Science*, **4**, 467–71.

Greenway, M. (1976). The grazing of *Thalassia testudinum* in Kingston Harbour, Jamaica. *Aquatic Botany*, **2**, 117–26.

Happ, G., Gosselink, J. G. and Davy, J. W. (1977). The seasonal distribution of organic carbon in a Louisiana estuary. *Estuarine and Coastal Marine Science*, **5**, 695–705.

Harrison, P. G. (1987). Natural expansion and experimental manipulation of seagrass (*Zostera* spp.) abundance and the response of infaunal invertebrates. *Estuarine, Coastal and Shelf Science*, **24**, 799–812.

Heald, E. J. (1969). *The Production of Organic Detritus in a S. Florida Estuary*. Ph.D. Thesis, University of Miami, 110 pp.

Heinle, D. R. and Flemer, D. A. (1976). Flows of materials between poorly flooded tidal marshes and an estuary. *Marine Biology*, **35**, 359–73.

Hutchings, P. and Saeger, P. (1987). *Ecology of Mangroves*. University of Queensland Press, St Lucia, Queensland, Australia.

Jackson, R. H., Williams, P. J. le B. and Joint, I. R. (1987). Freshwater phytoplankton in the low salinity region of the River Tamar estuary. *Estuarine, Coastal and Shelf Science*, **25**, 299–311.

Jefferies, R. L. (1972). Aspects of salt-marsh ecology with particular reference to inorganic plant nutrition. In R. S. K. Barnes and J. Green, (eds.), *The Estuarine Environment*. Applied Science Publishers, London, 61–85.

Joint, I. R. (1978). Microbial production of an estuarine mudflat. *Estuarine and Coastal Marine Science*, **7**, 185–95.

Khafji, A. K. and Norton, T. A. (1979). The effects of salinity on the distribution of *Fucus ceranoides*. *Estuarine and Coastal Marine Science*, **8**, 433–9.

Kirby-Smith, W. W. (1976). The detritus problem and the feeding and digestion of an estuarine organism. In M. Wiley (ed.), *Estuarine Processes (I)*. Academic Press, London, 469–79.

Laane, R. W. P. M. (1982). Chemical characteristics of the organic matter in the water-phase of the Ems-Dollard estuary. *Biologisch onderzoek eems-dollard estuarium*, **6**, 1–134.

Lively, J. S., Kaufman, Z. and Carpenter, E. J. (1983). Phytoplankton ecology of a barrier island estuary: Great South Bay, New York. *Estuarine, Coastal and Shelf Science*, **16**, 51–68.

Long, S. P. and Mason, C. F. (1983). *Saltmarsh Ecology* Blackie, Glasgow, 168 pp.

Macintyre, H. L., Geider, R. J. and Miller, D. C. (1996). Microphytobenthos: The ecological role of the "secret garden" of unvegetated, shallow-water marine habitats, 1. *Estuaries*, **19(2A)**, 186–201.

Malone, T. C. (1977). Environmental regulation of phytoplankton productivity in the Lower Hudson estuary. *Estuarine and Coastal Marine Science*, **5**, 157–71.

McLusky, D. S. (2001). North Sea estuaries. *Senckenbergia maritima*, **31**, 177–86.

McRoy, C. P. and Helfferich, C. (eds.) (1977). *Seagrass Ecosystems: A Scientific Perspective.* Marcel Dekker, New York, 314 pp.

Odum, E. P. and de la Cruz, A. A. (1967). Particulate organic detritus in a Georgia salt marsh/estuarine ecosystem. In Lauff, G. H. (ed.) *Estuaries*, American Association for the Advancement of Science, Washington, D.C., **83**, 383–94.

Oviatt, C., Keller, A., and Reed, L. (2002). Annual primary production in Narragansett Bay with no Bay-wide Winter-Spring phytoplankton bloom. *Estuarine Coastal and Shelf Science*, **54**, 1013–26

Pickval, J. C. and Odum, W. E. (1977). Benthic detritus in a salt marsh tidal creek. In M.Wiley (ed.), *Estuarine Processes (II)*. Academic Press, London, 280–92.

Pierce, S. M. (1983). Estimation of the non-seasonal production of *Spartina maritima* in a South African estuary. *Estuarine, Coastal and Shelf Science*, **16**, 241–54.

Pinckney, J. L., Paerl, H. W., Tester, P. and Richardson, T. L. (2001). The role of nutrient loading and eutrophication in estuarine ecology. *Environmental Health Perspectives*, **109**, (Suppl 5), 699–706.

Pomeroy, L. R. *et al.* (1977). Flow of organic matter through a salt marsh. In M. Wiley (ed.), *Estuarine Processes (II)*. Academic Press, London, 270–9.

Postma, H. (1988). Tidal flat areas. In B-O Jansson (ed.), *Coastal-Offshore Ecosystem Interactions.* Springer-Verlag, Berlin, 102–21.

Raffaelli, D. G., Raven, J. R. and Poole, L. J. (1998). Ecological impact of green macroalgal blooms. *Oceanography and Marine Biology; An Annual Review*, **36**, 97– 126.

Rasmussen, E. (1973). Systematics and ecology of the Isefjord marine fauna. *Ophelia*, **11**, 1–495.

Robertson, A. I. (1988). Decomposition of mangrove leaf litter in tropical Australia. *Journal of Experimental Marine Biology and Ecology*, **116**, 235–47.

Sand-Jensen, K. (1975). Biomass, net production and growth dynamics in an eelgrass (*Zostera marina*) population in Vellerup Vej, Denmark. *Ophelia*, **14**, 185–201.

Strachal, G. and Ganning, B. (1977). *Boken om havet.* Forskning och Framsteg, Stockholm, 132 pp. (in Swedish).

Teal, J. M. (1962). Energy flow in the salt marsh ecosystem of Georgia. *Ecology*, **43**, 614–62.

Tenore, K. R. (1981). Organic nitrogen and caloric content of detritus (I). *Estuarine, Coastal and Shelf Science*, **12**, 39–47.

Underwood, G. J. C. and Kromkamp, J. (1999). Primary production by phytoplankton and microphytobenthos in estuaries. *Advances in Ecological Research*, **29**, 93–153.

Valiela, I. and Teal, J. M. (1979). The nitrogen budget of a salt marsh ecosystem *Nature*, **280**, 652–6.

Van Es, F. B. (1977). A preliminary carbon budget for a part of the Ems estuary: The Dollard. *Helgolander Wiss. Meeresunters*, **30**, 283–291.

Van Raalte, C. D., Valiela, I. and Teal, J. M. (1976). Production of epibenthic salt marsh algae: light and nutrient limitation. *Limnology and Oceanography*, **21**, 862–72.

Van Valkenberg, S. D. *et al.* (1978). A comparison by size class and volume of detritus versus phytoplankton in Chesapeake Bay. *Estuarine and Coastal Marine Science*, **6**, 569–82.

Williams, P. J. le B. (1981). Primary productivity and heterotrophic activity in estuaries. In *River Inputs to Ocean Systems*, United Nations, New York, 243–58.

Williams, R. B. (1972). Annual phytoplanktonic production in a system of shallow temperate estuaries. In H.Barnes (ed.), *Some Contemporary Studies in Marine Science*. Aberdeen University Press, Aberdeen, 699–716.

Chapter Four

Ankar, S. and Elmgren, R. (1976). The benthic macro- and meio-fauna of the Askö Landsort Area (Northern Baltic proper). A stratified random sampling survey. *Contributions from the Asko Laboratory, University of Stockholm*, **11**, 1–115.

Austen, M. C. and Warwick, R. M. (1995). Effects of manipulation of food supply on estuarine meiobenthos. *Hydrobiologia*, **311**, 175–84.

Baird, D. and Ulanowicz, R. E. (1993). Comparative study of the trophic structure, cycling and ecosystem properties of four tidal estuaries. *Marine Ecology Progress Series*, **99**, 221–37.

Beukema, J. J. (1976). Biomass and species richness of the macro-benthic animals living on the tidal flats of the Dutch Wadden Sea. *Netherlands Journal of Sea Research*, **10**, 236–61.

Beukema, J. J., Cadee, G. C. and Hummel, H. (1983). Differential variability in time and space of numbers in suspension feeding and deposit feeding benthic species in a tidal flat area. *Oceanologica Acta*, special volume, 21–6.

Beukema, J. J., Knol, E. and Cadee, G. C. (1985). Effects of temperature on the length of the annual growing season in the tellinid bivalve *Macoma balthica* living on tidal flats in the Dutch Wadden Sea. *Journal of Experimental Marine Biology and Ecology*, **90**, 129–44.

Bloom, S. A., Simon, J. L. and Hunter, V. D. (1972). Animal-sediment relations and community analysis of a Florida estuary. *Marine Biology*, **13**, 43–56.

Buhr, K. J. and Winter, J. E. (1977). Distribution and maintenance of a *Lanice conchilega* association in the Weser estuary, with special reference to the suspension-feeding behaviour of *Lanice conchilega*. In B. Keegan *et al.* (eds.), *Proceedings of the 11th Europ. Mar. Biol. Symp.* Pergamon Press, Oxford, 101–13.

Coull, B. C. (ed.) (1977). *Ecology of Marine Benthos.* University of South Carolina Press, Columbia, S.C., Belle W. Baruch Library in Marine Science, **6**, 1–467.

Cranford, P. J., Peer, D. L. and Gordon, D. C. (1985). Population dynamics and production of *Macoma balthica* in Cumberland basin and Shepody Bay, Bay of Fundy. *Netherlands Journal of Sea Research*, **19**, 135–46.

Dame, R. F. (1976). Energy flow in an intertidal oyster population. *Estuarine and Coastal Marine Science*, **4**, 243–53.

De Wilde, P. A. W. J. (1975). Influence of temperature on behaviour, energy metabolism and growth of *Macoma balthica*. In H. Barnes (ed.), *Proceedings of 9th Europ. Mar. Biol. Symp.* Aberdeen Univ. Press, 239–59.

Elliott, M. and Kingston, P.F. (1987). The sublittoral benthic fauna of the Estuary and Firth of Forth. *Proceedings of the Royal Society of Edinburgh*, **93B**, 449–65.

Elliott, M. and O'Reilly, M.G. (1991). The variability and prediction of marine benthic community parameters. In M. Elliott and J-P Ducrotoy (eds.), *Estuaries and Coasts: Spatial and Temporal Intercomparisons*. Olsen and Olsen, Fredensborg, Denmark, 231–38.

Elliott, M. and Taylor, C. J. L. (1989). The production ecology of the subtidal benthos of the Forth Estuary, Scotland. *Scientia Marina*, **53 (2–3)**, 531–41.

Elmgren, R. (1984). Trophic dynamics in the enclosed, brackish Baltic Sea. *Rapp. P-v. Reun. Cons. int. Explor. Mer.*, **183**, 152–69.

Froneman, P. W. (2000). Feeding studies on selected zooplankton in a temperate estuary, South Africa. *Estuarine Coastal and Shelf Science*, **51**, 543–52

Froneman, P. W. (2001). Seasonal changes in zooplankton biomass and grazing in a temperate estuary, South Africa. *Estuarine Coastal and Shelf Sciences*, **52**, 543–54.

Giere, O. and Pfannkuche, O. (1982). Biology and ecology of marine oligochaeta, a review. *Oceanography and Marine Biology, Annual Review*, **20**, 173–308.

Gilbert, M. A. (1973). Growth rate, longevity and maximum size of *Macoma balthica. Biological Bulletin Marine Biological Laboratory, Wood's Hole*, **145**, 119–26.

Gillet, P. and Torresani, S. (2002). Structure of the population and secondary production of *Hediste diversicolor* in the Loire estuary, Atlantic coast, France. *Estuarine, Coastal and Shelf Science*, **56**, 621–8.

Heip, C. and Herman, R. (1979). Production of *Nereis diversicolor* O.F. Müller (Polychaeta) in shallow brackish-water pond. *Estuarine and Coastal Marine Science*, **8**, 297–305.

Herman, P. M. J. and Scholten, H. (1990). Can suspension feeders stabilise estuarine ecosystems? In M Barnes and R N Gibson (eds.), *Trophic Relationships in the Marine Environment*. pp 104–16. Aberdeen University Press.

Herman, P. M. J., Middleburg, J. J., Van De Koppel, J. and Heip, C. H. R. (1999). Ecology of estuarine macrobenthos. *Advances in Ecological Research*, **29**, 195–240

Hummel, H. (1985). An energy budget for a *Macoma balthica* (Mollusca) population living on a tidal flat in the Dutch Wadden Sea (I and II).

Netherlands Journal of Sea Research, **19**, 52–83 and 84–92.

Hylleberg, J. (1975). Selective feeding by *Abarenicola pacflca* with notes on *Arabenicola vagabunda* and a concept of gardening in lugworms. *Ophelia*, **14**, 113–37.

Jones, N. V. (1988). Life in the Humber, invertebrate animals. In N. V. Jones (ed.), *A Dynamic Estuary. Man, Nature and the Humber*. Hull University Press, 58–70.

Kruger, F. (1971). Bau und leben des wattwurmes *Arenicola marina*. *Helgolander wiss. Meeresunters.* **22**, 149–200.

Kuipers, B. R., De Wilde, P. A. W. J. and Creutberg, F. (1981). Energy flow in a tidal flat ecosystem. *Marine Ecology, Progress Series*, **5**, 215–21.

Lee, W. Y. and McAlice, B. J. (1979). Sampling variability of marine zooplankton in a tidal estuary. *Estuarine and Coastal Marine Science*, **8**, 565–82.

Madsen, P. B. and Jensen, K. (1987). Population dynamics of *Macoma balthica* in the Danish Wadden Sea in an organically enriched area. *Ophelia*, **27**, 197–208.

Mazik, K. and Elliott, M. (2000). The effects of chemical pollution on the bioturbation potential of estuarine intertidal mudflats. *Helgoland Marine Research*, **54**, 99–109

McLusky, D. S. (1987). Intertidal habitats and benthic macrofauna of the Forth estuary. *Proceedings of the Royal Society of Edinburgh*, **93B**, 389–400.

McLusky, D.S., Elliott, M. and Warnes, J. (1978). The impact of pollution on the intertidal fauna of the estuarine Firth of Forth. In D. S. McLusky and A. J. Berry (eds.), *Proceedings of the 12th Europ. Mar. Biol. Symposium.*, Pergamon Press, Oxford, 203–10.

McLusky, D. S. and Elliott, M. (1981). The feeding and survival strategies of estuarine molluscs. In N. V. Jones and W. J. Wolff (eds.), *Feeding and Survival Strategies in Estuarine Organisms*. Plenum Press, New York, 109–22.

McLusky, D.S. and McIntyre, A. D. (1988). Characteristics of the benthic fauna. In H. Postma and J. J. Zijlstra (eds.), *Continental Shelves*. Elsevier, Amsterdam, 131–54.

McLusky, D. S., Hull, S.C. and Elliott, M. (1993). Variations in the intertidal and subtidal macrofauna and sediments along a salinity gradient in the Upper Forth estuary. *Netherlands Journal of Aquatic Ecology*, **27**, 101–9.

Mees, J. and Jones, M. B. (1997). The hyperbenthos. *Oceanography and Marine Biology: An Annual Review*, **35**, 221–56.

Meire, P. M., Seys, J., Buijs, J. and Coosen, J. (1994). Spatial and temporal patterns of intertidal macrobenthic populations in the Oosterschelde – are they influenced by the construction of the storm-surge barrier? *Hydrobiologia*, **283**, 152–82.

Miller, C. B. (1983). The zooplankton of estuaries. In B. H. Ketchum (ed.), *Estuaries and Enclosed Seas*. Elsevier, Amsterdam, 103–49.

Møller, P. and Rosenberg, R. (1982). Production and abundance of the amphipod *Corophium volutator* on the west coast of Sweden. *Netherlands Journal of Sea Research*, **16**, 127–40.

Møller, P. and Rosenberg, R. (1983). Recruitment, abundance and production of *Mya arenaria* and *Cardium edule* in marine shallow waters, western Sweden. *Ophelia*, **22**, 33–55.

Moore, C. G. (1987). Meiofauna of the industrialised estuary and Firth of Forth, Scotland. *Proceedings of the Royal Society of Edinburgh*, **93B**, 415–30.

Muus, B. J. (1967). The fauna of Danish estuaries and lagoons. Distribution and ecology of dominating species in the shallow reaches of the mesohaline zone. *Medd. fra. Danmarks fiskeri og Havundersogelser. Ny serie*, **5**, 1–315.

Nicholls, D. J., Tubbs, C. R. and Haynes, F. N. (1981).The effect of green algal mats on intertidal macrobenthic communities and their predators. *Kieler Meeresforsch.* **5**, 511–20.

Nielsen, M. V. and Kofoed, L. H. (1982). Selective feeding and epipsammic browsing by the deposit-feeding amphipod *Corophium volutator*. *Marine Ecology Progress Series*, **10**, 81–8.

Park, G. S. and Marshall, H. G. (2000). The trophic contribution of rotifers in tidal freshwater and estuarine habitats. *Estuarine Coastal and Shelf Sciences*, **51**, 729–42.

Peer, D. L., Linkletter, L. E. and Hicklin, P. W. (1986). Life history and reproductive biology of *Corophium volutator*. *Netherlands Journal of Sea Research*, **20**, 359–73.

Revelante, N. and Gilmartin, M. (1987). Seasonal cycle of the ciliated protozoan and micrometazoan biomass in a Gulf of Maine estuary. *Estuarine, Coastal and Shelf Science*, **25**, 581–98.

Roddie, B. D., Leakey, R. J. G. and Berry, A. J. (1984). Salinity-temperature tolerance and osmoregulation in *Eurytemora affinis*, in relation to its distribution in the zooplankton of the upper reaches of the Forth estuary. *Journal of experimental marine biology and ecology*, **79**, 191–211.

Rosenberg, R., Nilsson, H. C. and Diaz, R. J. (2001). Responses of benthic fauna and changing sediment redox profiles over a hypoxic gradient. *Estuarine Coastal and Shelf Sciences*, **53**, 343–50.

Sandifer, P. A. (1975). The role of pelagic larvae in recruitment to populations of decapod crustacea in the York River estuary and adjacent Chesapeake bay. *Estuarine and Coastal Marine Science*, **3**, 269–80.

Smaal, A. C. and Prins, T. C. (1993). The uptake of organic matter and the release of inorganic nutrients by bivalve suspension feeder beds. In R. F. Dame (ed.), *Bivalve Filter Feeders in Estuarine and Coastal Ecosystem Processes*. NATO ASI Series G, **33**, 271–98.

Taylor, C. J. L. (1987). The zooplankton of the Forth, Scotland. *Proceedings of the Royal Society of Edinburgh*, **93B**, 377–88.

Taylor, C. J. L. (1993). The zooplankton of the Forth Estuary. *Netherlands Journal of Aquatic Ecology*, **27 (2–4)**, 87–99.

Warwick, R. M. (1971). Nematode associations in the Exe estuary. *Journal of the Marine Biological Association, UK*, **51**, 439–54.

Warwick, R. M. (1981). Survival strategies of meiofauna. In N. V. Jones and W. J. Wolff (eds.), *Feeding and Survival Strategies of Estuarine Organisms*. Plenum Press, New York, 39–52.

Warwick, R. M., Joint, I. R. and Radford, P. J. (1979). Secondary production of the benthos in an estuarine environment. In R. L. Jefferies and A. J. Davy (eds.), *Ecological Processes in Coastal Environments*. Blackwell, Oxford, 429–50.

Widdows, J., Brinsley, M. D., Salkeld, P. N. and Elliott, M. (1998). Use of annular flumes to determine the influence of current velocity and bivalves on material fluxes at the sediment-water interface. *Estuaries*, **21(4A)**, 552–9.

Widdows, J., Brinsley, M. D., and Elliott, M. (1998). Use of in situ flume to quantify particle flux (biodeposition rates and sediment erosion) for an intertidal mudflat in relation to changes in current velocity and benthic macrofauna. In K. S. Black, D. M. Patterson and A. Cramp (eds.), *Sedimentary Processes in the Intertidal Zone*. Geological Society, London, Special Publications, **139**, 85–97.

Widdows, J., Brown, S., Brinsley, M. D., Salkeld, P. N. and Elliott, M. (2000). Temporal changes in intertidal sediment erodability: influence of biological and climatic factors. *Continental Shelf Research*, **20**, 1275–1289.

Wolff, W. J. (1973). The estuary as a habitat. An analysis of data on the soft-bottom macrofauna of the estuarine area of the Rivers Rhine, Meuse and Scheldt. *Zoologische Verhandelingen*, **126**, 1–242.

Wolff, W. J. (1977). A benthic food budget for the Grevelingen estuary, The Netherlands, and a consideration of the mechanisms causing high benthic secondary production in estuaries. In B. C. Coull (ed.), *Ecology of Marine Benthos*. University of South Carolina Press, Columbia, S.C., 267–80.

Wolff, W. J. (1983). Estuarine benthos. In B. H. Ketchum (ed.), *Estuaries and Enclosed Seas*. Elsevier, Amsterdam, 151–82.

Wooldridge, T. and Erasmus. T. (1980). Utilisation of tidal currents by estuarine zooplankton. *Estuarine and Coastal Marine Science*, **11**, 107–14.

Ysebaert, T. *et al.* (2000). Waterbird communities along the estuarine salinity gradient of the Schelde estuary. *Biodiversity and Conservation*, **9**, 1275–1296.

Ysebaert, T., Meire, P., Herman, P. M. J. and Verbeek, H. (2002). Macrobenthic species response surfaces along estuarine gradients: prediction by logistic regression. *Marine Ecology Progress Series*, **225**, 79–95.

Ysebaert, T., Herman, P. M. J., Meire, P., Craeymeersch, J. and Verbeek, H. (2003). Large scale spatial patterns in estuaries: estuarine macrobenthic communities in the Schelde estuary. *Estuarine, Coastal and Shelf Science*, **57**, 335–56.

Chapter Five

Ambrose, W. G. Jr. (1986). Estimate of removal rate of *Nereis virens* from an intertidal mudflat by gulls (*Larus* spp.). *Marine Biology*, **90**, 243–47.

Baird, D. and Milne, H. (1981). Energy flow in the Ythan estuary, Aberdeenshire, Scotland. *Estuarine, Coastal and Shelf Science*, **13**, 455–72.

Baran, E. (1999). A review of quantified relationships between mangroves and coastal resources. *Phuket Marine Biological Centre, Research Bulletin*, **62**, 57–64.

Berry, A. J. (1988). Annual cycle in *Retusa obtusa* of reproduction, growth and predation upon *Hydrobia ulvae*. *Journal of Experimental Marine Biology and Ecology*, **117**, 197–209.

Blaber, S. J. M. (1997). *Fish and Fisheries of Tropical Estuaries*. Chapman and Hall, London.

Blaber, S. J. M. (2000). *Tropical Estuarine Fishes: Ecology, Exploitation and Conservation*. Blackwell, Oxford.

Blaber, S. J. M. (2002). Fish in hot water: The challenges facing fish and fisheries research in tropical estuaries. *Journal of Fish Biology*, **61**, (Suppl. A), 1–20.

Boddeke, R. *et al.* (1986). Food availability and predator presence in a coastal nursery area of the brown shrimp *(Crangon crangon)*. *Ophelia*, **26**, 77–90.

Bryant, D. M. (1979). Effects of prey density and site character on estuary usage by over-wintering waders. *Estuarine and Coastal Marine Science*, **9**, 369–84.

Bryant, D. M. and Leng, J. (1976). Feeding distribution and behaviour of Shelduck in relation to food supply. *Wildfowl*, **27**, 20–30.

Bryant, D. M. and McLusky, D. S. (1997). Long term changes amongst wintering waders (Charadrii) on the Forth estuary: Effects of changing food densities. *Coastal Zone Topics*, **3**, 171–81

Claridge, P. *et al.* (1986). Seasonal changes in movements, abundance, size composition and diversity of the fish fauna of the Severn estuary. *Journal of the Marine Biological Association of the UK*, **66**, 229–58.

Costa, M. J. and Elliott, M. (1991). Fish usage and feeding in two industrialised estuaries – the Tagus, Portugal, and the Forth, Scotland. In M. Elliott and J.-P. Ducrotoy (eds.), *Estuaries and Coasts: Spatial and Temporal Intercomparisons*. Olsen and Olsen, Fredensborg, Denmark, 289–97.

Costanza, R. (1992). Towards an operational definition of ecosystem health. In R. Costanza, B. G. Norton and B. D. Haskell (eds.), *Ecosystem Health: New Goals for Environmental Management*. Island Press, Washington, D.C., 239–56.

Daan, N., Bromley, P. J., Hislop, J. R. G. and Nielsen, N. A. (1990). Ecology of North Sea fish. *Netherlands Journal of Sea Research*, **26**, 343–86.

Evans, P. R. *et al.* (1979). Short-term effects of reclamation of part of Seal Sands, Teesmouth, on wintering waders and Shelduck. *Oecologia*, **41**, 183–206.

Elliott, M. and Taylor, C. J. L. (1989). The structure and functioning of an estuarine/marine fish community in the Forth Estuary, Scotland. *Proceedings of the 21st European Marine Biology Symposium*, Gdansk, September, 1986. Polish Academy of Sciences – Institute of Oceanology, 227–40.

Elliott, M., O'Reilly, M. G. and Taylor, C. J. L. (1990). The Forth Estuary: A nursery and overwintering area for North Sea fish species. *Hydrobiologia*, **195**, 89–103.

Elliott, M. and Dewailly, F. (1995). Structure and components of European estuarine fish assemblages. *Netherlands Journal of Aquatic Ecology*, **29**, 397–417.

Elliott, M. and Hemingway, K. (eds.) (2002). *Fishes in Estuaries*. Blackwell Science, Oxford. 636 pp.

Gee, J. M. *et al.* (1985). Field experiments on the role of epibenthic predators in determining prey densities in an estuarine mudflat. *Estuarine, Coastal and Shelf Science*, **21**, 429–48.

Goss-Custard, J. D. *et al.* (1977). The density of migratory and overwintering Redshank and Curlew in relation to the density of their prey in SE England. *Estuarine and Coastal Marine Science*, **5**, 497–510.

Goss-Custard, J. D. and Moser, M. E. (1988). Rates of change in the numbers of Dunlin, *Calidris alpina*, wintering in British estuaries in relation to the spread of *Spartina anglic*. *Journal of Applied Ecology*, **25**, 95–109.

Gunter, G. (1967). Some relationships of estuaries to the fisheries of the Gulf of Mexico. In G. H. Lauff (ed.), *Estuaries*. American Association for the Advancement of Science, Washington, D. C., **83**, 621–38.

Henderson, P. *et al.* (1992). Trophic structure within the Bristol Channel: seasonality and stability in Bridgewater Bay. *Journal of the Marine Biological Association*, U.K. **72**, 675–90.

Hoff, J. G. and Ibara, R. M. (1977). Factors affecting the seasonal abundance, composition and diversity of fishes in a SE New England estuary. *Estuarine and Coastal Marine Science*, **5**, 665–78

Jager, Z. (1999). Selective tidal stream transport of Flounder larvae (*Platichthys flesus*) in the Dollard (Ems Estuary). *Estuarine Coastal and Shelf Science*, **49**, 347–62.

Jayamanne, S. and McLusky, D. S. (1997). Biology of the shrimps in the Forth estuary. *Coastal Zone Topics*, **3**, 147–56.

Kneib, R. T. (1997). The role of tidal marshes in the ecology of estuarine nekton. *Oceanography and Marine Biology: An Annual Review*, **35**, 163–220.

Kremer, P. (1977). Population dynamics and ecological energetics of a pulsed zooplankton predator, the Ctenophore *Mnemiopsis leidyi*. In M. Wiley (ed.), *Estuarine Processes (I)*. Academic Press, London, 197–205.

Lafaille, P., Feunteun, E., and Lefuvre, J.-C. (2000). Composition of fish communities in a European macrotidal salt marsh (the Mont Saint-Michel bay, France). *Estuarine Coastal and Shelf Sciences*, **51**, 429–38.

Lenanton, R. C. J. and Hodgkin, E. P. (1985). Life history strategies of fish in some temperate Australian estuaries. In A. Yáñez-Arancibia (ed.), *Fish Community Ecology in Estuaries and Coastal Lagoons: Towards an Ecosystem Integration*. UNAM Press, Mexico, 277–84.

Maes, J., Van Damme, P. A., Taillieu, A. and Ollevier, F. (1998). Fish communities along an oxygen-poor salinity gradient (Zeeschelde estuary, Belgium). *Journal of Fish Biology*, **52**, 534–46.

Marshall, S. and Elliott, M. (1996). The structure of the fish assemblage in the Humber Estuary, UK. *Publicaciones Especiales Instituto Espanol de Oceanografia*, **21**, 231–42.

Marshall, S. and Elliott, M. (1998). Environmental influences on the fish assemblage of the Humber estuary, UK. *Estuarine, Coastal and Shelf Science*, **46**, 175–84.

Mathieson, S., Cattrijsse, A., Costa, M. J., Drake, P., Elliott, M., Gardner, J. and Marchand, J. (2000). Fish assemblages of European tidal marshes: a functional guilds-based comparison. *Marine Ecology Progress Series*, **204**, 225–42.

Meredith, W. H. and Lotrich, V. A. (1979). Production dynamics of a tidal creek population of *Fundulus heteroclitus (L)*. *Estuarine and Coastal Marine Science*, **8**, 99–118.

Methven, D. A., Haedich, R. L. and Rose, G. A. (2001). The fish assemblage of a Newfoundland estuary: Diel, monthly and annual variation. *Estuarine Coastal and Shelf Science*, **52**, 669–88.

Pihl, L. (1982). Food intake of young cod and flounder in a shallow bay on the Swedish west coast. *Netherlands Journal of Sea Research*, **15**, 419–32.

Pihl, L. (1985). Food selection and consumption of mobile epibenthic fauna in shallow marine areas. *Marine Ecology, Progress Series*, **22**, 169–79.

Pihl, L. and Rosenberg, R. (1984). Food selection and consumption of the shrimp *Crangon crangon* in some shallow marine areas in western Sweden. *Marine Ecology, Progress Series*, **15**, 159–68.

Prater, A. J. (1981). *Estuary birds of Britain and Ireland*. T. and A. D. Poyser, Berkhamsted, 440 pp.

Pomfret, J. R., Elliott, M., O'Reilly, M. G. and Phillips, S. (1991). Temporal and spatial patterns in the fish communities of two North Sea estuaries. In M. Elliott and J.-P. Ducrotoy (eds.). *Estuaries and Coasts: Spatial and Temporal Intercomparisons*. Olsen and Olsen, Fredensborg, Denmark, 277–4.

Power, M. *et al.* (2000). Temporal abundance patterns and growth of juvenile herring and sprat from the Thames estuary 1977–1992. *Journal of Fish Biology*, **56**, 1408–26.

Raffaelli, D. and Milne, H. (1987). An experimental investigation of the effects of shorebird and flatfish predation on estuarine invertebrates. *Estuarine, Coastal and Shelf Science*, **24**, 1–13.

Raffaelli, D. and Hall, S. J. (1992). Compartments and predation in an estuarine food web. *Journal of Animal Ecology*, **61**, 551–60.

Reading, C. J. and McGrorty, S. (1978). Seasonal variations in the burying depth of *Macoma balthica* and its accessibility to wading birds. *Estuarine and Coastal Marine Science*, **6**, 135–44.

Ronn, C., Bonsdorff, E. and Nelson, W. G. (1988). Predation as a mechanism of interference within infauna in shallow brackish water soft bottoms; experiments with an infauna predator, *Nereis diversicolor* (O.F. Müller). *Journal of Experimental Marine Biology and Ecology*, **116**, 143–57.

Ross, S. T. (1986). Resource partitioning in fish assemblages; a review of field studies. *Copeia*, **2**, 352–88.

Sherman, K. and Duda, A. M. (1999). An ecosystem approach to global assessment and management of coastal waters. *Marine Ecology Progress Series*, **190**, 271–87.

Thormann, S. (1986). Seasonal colonisation and effects of salinity and temperature on species richness and abundance of fish of some brackish and estuarine shallow waters in Sweden. *Holoarctic Ecology*, **9**, 126–32.

Thormann, S. and Widerholm, A.-M. (1986). Food, habitat and time niches in a coastal fish species assemblage in a brackish water bay in the Bothnian Sea, Sweden. *Journal of Experimental Marine Biology and Ecology*, **95**, 67–86.

Whitfield, A. K. (1994). An estuary-association classification for the fishes of southern Africa. *South African Journal of Science*, **90**, 411–17.

Whitfield, A. K. (1998). Biology and ecology of fishes in Southern African estuaries. *Ichthyological*

Monographs of the J.L.B. Smith Institute of Ichthyology, **2**, 1–223.

Wolff, W. J., Mandos, M. A. and Sandee, A. J. J. (1981). Tidal migration of plaice and flounders as a feeding strategy. In N. V. Jones and W. J. Wolff (eds.), *Feeding and Survival Strategies in Estuarine Organisms*. Plenum Press, New York, 159–71.

Zwarts, L. (1986). Burying depth of the benthic bivalve *Scrobicularia plana* (da Costa) in relation to siphon-cropping. *Journal of Experimental Marine Biology and Ecology*, **101**, 25–39.

Chapter Six

Bayne, B. L. *et al.* (1985). *The Effects of Stress and Pollution on Marine Animals*. Praeger Publishers, New York, 315 pp.

Billen, G., Garnier, J., Deligne, C. and Billen, C. (1999). Estimates of early industrial inputs of nutrients to river systems: implications for coastal eutrophication. *Science of the Total Environment*, **243/244**, 43–52.

Blaber, S. J. M., Albaret, J.-J., Chong Ving Ching, Cyrus, D. P., Day, J. W., Elliott, M., Fonseca, D., Hoss, J., Orensanz, J., Potter, I. C. and Silvert, W. (2000). Effects of fishing on the structure and functioning of estuarine and nearshore ecosystems. *ICES Journal of Marine Science*, **57**, 590–602.

Blowers, A. and Hinchcliffe, S. (eds.) (2003). *Environmental Responses*. John Wiley and Sons, Chichester and Open University, Milton Keynes, 304 pp.

Bouwman, L. A., Romeijn, K. and Admiraal, W. (1984). On the ecology of meiofauna in an organically polluted estuarine mudflat. *Estuarine, Coastal and Shelf Science*, **19**, 633–53.

Bricker, S. B., Clement, C. G., Pirhalla, D. E., Orlando, S. P., and Farrow, D. R. G. (1999). *National Estuarine Eutrophication Assessment: Effects of Nutrient Enrichment in the Nations Estuaries*. NOAA, National Ocean Service, Special Projects Office and the National Centres for Coastal Ocean Science. 71 pp.

Brown, J. R., Gowen, R. J. and McLusky, D. S. (1987). The effect of salmon farming on the benthos of a Scottish sea loch. *Journal of Experimental Marine Biology and Ecology*, **109**, 39–51.

Bryan, G. W. (1984). Pollution due to heavy metals and their compounds. *Marine Ecology*, **5**, 1289–431.

Bryant, V. *et al.* (1984). Effect of temperature and salinity on the toxicity of chromium to three estuarine invertebrates *(Corophium volutator, Macoma balthica, Nereis diversicolor)*. *Marine Ecology, Progress Series*, **20**, 137–49.

CONCAWE (1981). *A field guide to coastal oilspill control and clean-up technologies*. Report 9/81, CONCAWE, Brussels, (accessed via http://www.concawe.org) 112 pp.

Conley, D. J., Kaas, H., Mohlenberg, F., Rasmussen, B., and Windolf, J. (2000). Characteristics of Danish estuaries. *Estuaries*, **23**, 820–37.

Coombs, T. L. and George, S. G. (1978). Mechanisms of immobilisation and detoxication of metals in marine organisms. In D. S. McLusky and A. J. Berry (eds.), *Proceedings of the 12th European Marine Biology Symposium*. Pergamon Press, Oxford, 179–87.

Coughlan, J. (1979). Aspects of reclamation in Southampton Water. In B. Knights and A. J. Phillips (eds.), *Estuarine and coastal land reclamation and storage*. Saxon House, London, 99–124.

Davenport, J., *et al.* (2003). *Aquaculture: the ecological issues*. Blackwell Science, Oxford, 89 pp.

De Jonge, V. N and Elliott, M. (2001). Eutrophication In J. Steele, K. K. Turekian and S. Thorpe (eds.), *Encyclopedia of Ocean Sciences*, Volume 2. Academic Press, London, 852–70.

De Jonge, V. N., Elliott, E. and Orive, E. (2002). Causes, historical development, effects and future challenges of a common environmental problem: eutrophication. *Hydrobiologia*, **475/476**, 1–19.

Dicks, B. (ed.) (1989). *Ecological Impacts of the Oil Industry*. John Wiley, Chichester, 316 pp.

Elliott, M. (1994). The analysis of macrobenthic community data. *Marine Pollution Bulletin*, **28**, 62–4.

Elliott, M. (2003). Biological pollutants and biological pollution – an increasing cause for concern. *Marine Pollution Bulletin*, **46**, 275–80.

Elliott, M. and Boyes, S. J. (2002). *Trophic and Organic Status of the Humber Estuary – Phase 2*. Report to Yorkshire Water Services by the Institute of Estuarine and Coastal Studies, University of Hull. ZBB591-F-2002. http://www.kelda-group.com/environment/reports_pdfs/returning/finalywph2.pdf

Elliott, M. and Griffiths, A. H. (1986). Mercury contamination in components of an estuarine ecosystem. *Water Science and Technology*, **18**, 161–70.

Elliott, M., Griffiths, A. H. and Taylor, C. J. L. (1988). The role of fish studies in estuarine pollution assessment. *Journal of Fish Biology*, **33(A)**, 51–62.

Elliott, M. and Griffiths, A. H. (1988). Contamination and effects of hydrocarbons on the Forth ecosystem. *Proceedings of the Royal Society of Edinburgh*, **93B**, 327–42.

Elliott, M., Nedwell, S., Jones, N. V., Read, S., Cutts, N. D. and Hemingway, K. I. (1998). *Intertidal sand and mudflats and subtidal mobile sandbanks (Volume II). An overview of dynamic and sensitivity characteristics for conservation management of marine SACs*. Scottish Association for Marine Science, Oban, for the UK Marine SAC project, pp. 151 (accessed via http://www.english-nature.org.uk/uk-marine/reports/pdfs/sandmud.pdf).

Essink, K. (1978). The effects of pollution by organic waste on macrofauna in the Eastern Dutch Wadden Sea. *Netherlands Institute for Sea Research, Publication Series*, **1**, 1–135.

GESAMP (Group of Experts on Marine Pollution) (1990). *The State of the Marine Environment*. Blackwell Science Publ., London, 146 pp.

Gowen, R. *et al.* (1988). *Investigations into benthic enrichment, hypernutrification and eutrophication associated with mariculture in Scottish coastal waters*, Department of Biological Science, University of Stirling, 289 pp.

Hall, S. J. (1999). *The Effects of Fishing on Marine Ecosystems and Communities*. Blackwell Science, Oxford, 274 pp.

Harding, L. E. (1992). Measures of marine environmental quality. *Marine Pollution Bulletin*, **25**, 23–27.

Heap, J.M., Elliott, M. and Rheinalt, A. P. (1991). The marine disposal of sewage-sludge. *Ocean and Shoreline Management*, **16**, 291–312.

Hunt, G. J. (1988). *Radioactivity in surface and coastal waters of the British Isles, 1987*. Ministry of Agriculture, Fisheries and Food, Lowestoft, Suffolk, Aquatic Environment Monitoring Report, Number 19, 67 pp.

Jennings, S. and Kaiser, M. J. (1998). The effects of fishing on marine ecosystems. *Advances in Marine Biology*, **34**, 201–352.

Johnston, R. (1984). Oil pollution and its management. *Marine Ecology*, **5**, 1433–582.

La Jeunesse, I. and Elliott, M. (2004). Anthropogenic regulation of the phosphorus balance in the Thau catchment – coastal lagoon system (Mediterraean Sea, France) over 24 years. *Marine Pollution Bulletin*, **48(7–8)**, 679–87.

Laws, E. A. (1993). *Aquatic Pollution: An Introductory Text*, 2nd Edition. John Wiley, New York, 611 pp.

Langston, W. J. and Bebianno, M. J. (eds.) (1998). *Metal Metabolism in Aquatic Environments*. Chapman and Hall, London/Kluwer, Dordrecht, 448 pp.

Leatherland, T. M. (1987). Radioactivity in the Forth, Scotland. *Proceedings of the Royal Society of Edinburgh*, **93B**, 299–301.

Leppakoski, E. (1975). Assessment of degree of pollution on the basis of macrozoobenthos in marine and brackish water environments. *Acta Academiae Aboensis (B)*, **35**, 1–89.

MacKay, D. W. (1986). Sludge dumping in the Firth of Clyde-a containment site. *Marine Pollution Bulletin*, **17**, 91–5.

McLusky, D. S. (1982). The impact of petro-chemical effluent on the fauna of an intertidal estuarine mudflat. *Estuarine, Coastal and Shelf Science*, **14**, 489–499.

McLusky, D. S., Bryant, V. and Campbell, R. (1986). The effects of temperature and salinity on the toxicity of heavy metals to marine and estuarine invertebrates. *Oceanography and Marine Biology, Annual Review*, **24**, 481–520.

McLusky, D. S., Bryant, D. M. and Elliott, M. (1992). The impact of land-claim on the invertebrates, fish and birds of the Forth estuary. *Aquatic Conservation: Marine and Freshwater Ecosystems*, **2**, 211–22.

McManus, J. W. and Pauly, D. (1990). Measuring ecological stress: variations on a theme by R. M. Warwick. *Marine Biology*, **106**, 305–8.

MEMG (Marine Environmental Monitoring Group; Group Coordinating Sea Disposal Monitoring) (2003). *Final report of the Dredging and Dredged Material Disposal Monitoring Task-team*. Science Series, Aquatic Environment Monitoring Report, CEFAS, Lowestoft, UK, Number 55, 52 pp.

Moriarty, F. (1999). *Ecotoxicology: The Study of Pollutants in Ecosystems*. 3rd Edition, Academic Press, London, 347 pp.

Moore, D. C. and Davies, I. M. (1987). Monitoring the effects of the disposal at sea of Lothian Region sewage sludge. *Proceedings of the Royal Society of Edinburgh*, **93B**, 467–77.

Murby, P. (2001). *Intertidal change – the causes, extent and implications of a regional view from Eastern England*. Unpublished M.Sc. thesis, University of Hull, U.K., 227 pp.

National Research Council (2000). *Clean Coastal Waters – understanding and reducing the effects of nutrient pollution.* National Academy Press, Washington DC, 405 pp.

Nixon, S. W. (1997). Prehistoric nutrient inputs and productivity in Narragansett Bay. *Estuaries*, **20**, 253–61.

North Sea Task Force (1993). *North Sea Quality Status Report 1993.* Oslo and Paris Commissions, London, and Olsen and Olsen, Fredensborg, 132 pp.

Pearson, T. H., Ansell, A. D. and Robb, L. (1986). The benthos of the deeper sediments of the Firth of Clyde, with particular reference to organic enrichment. *Proceedings of the Royal Society of Edinburgh*, **90B**, 329–50.

Pearson, T. H. and Rosenberg, R. (1978). Macrobenthic succession in relation to organic enrichment and pollution of the marine environment. *Oceanography and Marine Biology: Annual Review*, **16**, 229–311.

Pearson, T. H. and Stanley, S. O. (1979). Comparative measurement of the Redox potential of marine sediments as a rapid means of assessing the effect of organic pollution. *Marine Biology*, **53**, 371–80.

Pillay, T. V. R. (1993). *Aquaculture: Principles and Practices.* Fishing News Books, Blackwells, Oxford, 575 pp.

Pillay, T. V. R. (2003). *Aquaculture and the Environment.* 2nd edition. Blackwells, Oxford, 189 pp.

Postma, H. (1985). Eutrophication of Dutch coastal waters. *Netherlands Journal of Zoology*, **35**, 348–59.

Rosenberg, R. (ed.) (1984). *Eutrophication in marine waters surrounding Sweden—a review.* Swedish National Environment Protection Board, Solna, Report No. 1808, 140 pp.

Rosenberg, R. (1985). Eutrophication—the future marine coastal nuisance? *Marine Pollution Bulletin*, **16**, 227–31.

Shaw, T. L. (ed.) (1980). *An Environmental Appraisal of Tidal Power Stations, with Particular Reference to the Severn Barrage.* Pitman, London.

Steimle, F. W. Jr. (1985). Biomass and estimated productivity of the benthic macrofauna in the New York Bight: A stressed coastal area. *Estuarine, Coastal and Shelf Science*, **21**, 539–54.

Tenore, K. R. *et al.* (1982). Coastal upwelling in the Rias Bajas, NW Spain: contrasting the benthic regions of the Rias de Arosa and de Muros. *Journal of Marine Research*, **40**, 701–72.

Tenore, K. R., Corral, J. and Gonzales, N. (1985). Effects of intense mussel culture on food chain patterns and production in coastal Galicia, NW Spain. *ICES C.M.*, 1985/F:62 Sess W.

Turnpenny, A. W. H. and Coughlan, J. (1992). Power generation on the British coast: thirty years of marine biological research. *Hydroecologie Appliquee*, **4(1)**, 1–12.

Van Impe, J. (1985). Estuarine pollution as a probable cause of increase of estuarine birds. *Marine Pollution Bulletin*, **16**, 271–6.

Van der Veer, H. W. *et al.* (1985). Dredging activities in the Dutch Wadden Sea: effects on macrobenthic infauna. *Netherlands Journal of Sea Research*, **19**, 183–90.

Warwick, R. M. (1986). A new method for detecting pollution effects on marine macrobenthic communities. *Marine Biology*, **92**, 557–62.

Wharfe, J. R., Wilson, S. R. and Dines, R. A. (1984). Observations of the fish populations of an east coast estuary. *Marine Pollution Bulletin*, **15**, 133–6.

Wilber, D. H. and Clarke, D. G. (2001). Biological effects of suspended solids: a review of suspended solids impacts on fish and shellfish with relation to dredging activities in estuaries. *North American Journal of Fisheries Management*, **21**, 855–75.

Zajac, R. N. and Whitlach, R. B. (1982). The responses of estuarine infauna to disturbance (I and II). *Marine Ecology, Progress Series*, **10**, 1–14, 15–27.

Willis, K. and Ling, N. (2003). The toxicity of emamectin benzoate, an aquaculture pesticide, to planktonic marine copepods. *Aquaculture*, **221**, 289–97.

Chapter Seven

Alawi, Y. S., McConkey, B. J., Dixon, D. G. and Greenberg, B. M. (2000). Measurement of short- and long-term toxicity of polycyclic aromatic hydrocarbons using luminescent bacteria. *Ecotoxicology and Environmental Safety*, **51**, 12–21.

Beardmore, J. A., Barker, C. J., Battaglia, B., Berry, R. J., Crosby-Longwell, A., Payne, J. F. and Rosenfield, A. (1980). The use of genetical approaches to monitoring biological effects of pollution. *Rapports et Procés-verbaux des Réunions du Conseil International pour l'Exploration de la Mer*, **179**, 299–305.

Begon, M., Harper, J. L. and Townsend, C.R. (1990). *Ecology – Individuals, Populations and Communities.* 2nd edition. Blackwell Science Publications, Oxford, 945 pp.

Blackstock, J. (1984). Biochemical metabolic regulatory responses of marine invertebrates to natural environmental change and marine pollution. *Oceanography and Marine Biology, Annual Review*, **22**, 263–313.

Borgmann, U. (2003). Derivation of cause-effect sediment quality guidelines. *Canadian Journal of Fisheries and Aquatic Science*, **60**, 352–60.

Calmano, W., Ahlf, W. and Förstner, U. (1996). Sediment quality assessment: chemical and biological approaches. In W. Calmano and U. Förstner (eds.), Sediments and toxic substances. *Environmental Effects and Ecotoxicity.* Springer-Verlag, Berlin, 1–35.

Chapman, P. M., Wang, F., Janssen, C., Persoone, G. and Allen, H. E. (1998). Ecotoxicology of metals in aquatic sediments: binding and release, bioavailability, risk assessment and remediation. *Canadian Journal of Fisheries and Aquatic Science*, **55**, 2221–43.

Davis, J. *et al.* (eds.) (2001). *Marine Monitoring Handbook.* Joint Nature Conservation Committee, Peterborough, UK, 405 pp.

Elliott, M. (1993). The Quality of Macrobiological Data. *Marine Pollution Bulletin*, **25**, 6–7.

Elliott, M. (1996). The derivation and value of ecological quality standards and objectives. *Marine Pollution Bulletin*, **32**, 762–3.

Elliott, M. (2002). The role of the DPSIR approach and conceptual models in marine environmental management: an example for offshore wind power. *Marine Pollution Bulletin*, **44(6)**, iii–vii.

Ellis, J. I., Schneider, D. C. and Thrush, S. F. (2000). Detecting anthropogenic disturbance in an environment with multiple gradients of physical disturbance, Manukua Harbour, New Zealand. *Hydrobiologia*, **440**, 379–91.

Engel, D. W. and Brouwer, M. (1984). Trace metal binding proteins in marine molluscs and crustaceans. *Marine Environmental Research*, **13**, 177–94.

Freitas, R., Rodrigues, A. M. and Quintino, V. (2003). Benthic biotopes remote sensing using acoustics. *Journal of Experimental Marine Biology and Ecology*, **285–286**, 339–53.

Gauch, H. G. (1982). *Multivariate Analysis in Community Ecology.* Cambridge University Press, Cambridge.

Gee, J. M. (1983). Chapter 7 Sampling and analysis of fish populations. In A. W. Morris (ed.). *Practical Procedures for Estuarine Studies.* Natural Environment Research Council, Plymouth, 213–38.

Gosling, E. M. (2003). *Bivalve Molluscs: Biology, Ecology and Culture.* Fishing News Books, Blackwells, Oxford. 443 pp.

Glasson, J., Therivael, R. and Chadwick, A. (1999). *Introduction to Environmental Assessment*, 2nd Edition. UCL Press, London. 496 pp.

Harrop, D. O. and Nixon, J. A. (1999). *Environmental Assessment in Practice.* Routledge, London.

Hinkle-Conn, C., Fleeger, J. W., Gregg, J. C. and Carman, K. R. (1998). Effects of sediment-bound polycyclic aromatic hydrocarbons on feeding behaviour in juvenile spot (*Leiostomus xanthurus* lacépède: Pisces). *Journal of Experimental Marine Biology and Ecology*, **227**, 113–32.

Hiscock, K. (ed.) (1996). *Marine Nature Conservation Review: Rationale and Methods.* Joint Nature Conservation Committee, Peterborough, UK, 167 pp.

Jongman, R. H. G., Ter Braak, C. J. F. and Van Tongeran, O. F. R. (1995). *Data Analysis in Community and Landscape Ecology*, 2nd Edition. Cambridge University Press, Cambridge, 299 pp.

Kirby, M. F. *et al.* (1999). Hepatic EROD activity in flounder (*Platichthys flesus*) as an indicator of contaminant exposure in English estuaries. *Marine Pollution Bulletin*, **38**, 676–86.

Krebs, C. J. (1998). *Ecological Methodology*, 2nd Edition. Longman, Harlow, Essex, UK, 620 pp.

Lam, P. K. S. and Gray, J. S. (2001). Predicting effects of toxic chemicals in the marine environment. *Marine Pollution Bulletin*, **42(3)**, 169–73.

Lindsay, D. M. and Sanders, J. G. (1990). Arsenic uptake and transfer in a simplified estuarine food chain. *Environmental Toxicology and Chemistry*, **9**, 391–5.

Long, E. R. and Chapman, P. M. (1985). A sediment quality triad: measures of sediment contamination, toxicity and infaunal community composition in Puget Sound. *Marine Pollution Bulletin*, **16**, 405–15.

Lotufo, G. R. (1997). Toxicity of sediment-associated PAHs to an estuarine copepod: effects on survival, feeding, reproduction and behaviour. *Marine Environmental Research*, **44**, 149–66.

Lotufo, G. R. and Fleeger, J. W. (1997). Effects of sediment-associated phenanthrene on survival, development and reproduction of two species of meiobenthic copepods. *Marine Ecology Progress Series*, **151**, 91–102.

Marshall, S. and Elliott, M. (1997). A comparison of univariate and multivariate numerical and graphical techniques for determining inter- and intra-specific feeding relationships in estuarine fish. *Journal of Fish Biology*, **51**, 526–45.

Office of the Deputy Prime Minister (2004). *Environmental Impact Assessment – A Guide to Procedures*. Thomas Telford, Tonbridge, Kent, 58 pp., accessed through http://www.odpm.gov.uk/stellent/groups/odpm_planning/documents/page/odpm_plan-026667.hcsp

Pascoe, D. (1983). *Toxicology*. Edward Arnold, London, 60 pp.

Rand, G. M. (ed.) (1995). *Fundamentals of Aquatic Toxicology: Effects, Environmental Fate and Risk Assessment*, 2nd Edition. Taylor and Francis, London, 1125 pp.

Smith, L. G. (1993). *Impact Assessment and Sustainable Resource Management*. Longman, Harlow, England, 210 pp.

Spellerberg, I. F. (1991). *Monitoring Environmental Change*. Cambridge University Press, Cambridge, 334 pp.

Stebbing, A. R. D., Deflethsen, V. and Carr, M. (1992). Biological effects of contaminants in the North Sea: results of the ICES/IOC Bremerhaven Workshop. *Marine Ecology Progress Series*, **91 (1–3)**, 1–361.

Thrush, S. F. and Roper, D. S. (1988). Merits of macrofaunal colonization of intertidal mudflats for pollution monitoring: a preliminary study. *Journal of Experimental Marine Biology and Ecology*, **116**, 219–33.

Underwood, A. J. and Peterson, C. H. (1988). Towards an ecological framework for investigating pollution. *Marine Ecology Progress Series*, **46**, 227–34.

Warwick, R. M. (1986). A new method for detecting pollution effects on marine macrobenthic communities. *Marine Biology*, **92**, 557–62.

Wathern, P. (ed.) (1988). *Environmental Impact Assessment*. Routledge, London, 332 pp.

White, H. H. (ed.) (1985). *Concepts in Marine Pollution Measurements*. Maryland Sea Grant Publ., University of Maryland, 734 pp.

Whitfield, A. K. and Elliott, M. (2002). Fishes as indicators of environmental and ecological changes within estuaries – a review of progress and some suggestions for the future. *Journal of Fish Biology*, **61(A)**, 229–50.

Woodhead, R. J., Law, R. J. and Matthiessen, P. (1999). Polycyclic aromatic hydrocarbons in surface sediments around England and Wales, and their possible biological significance. *Marine Pollution Bulletin*, **38**, 773–90.

Chapter Eight

By its nature, estuarine management and legislation is a rapidly and constantly changing field. Consequently, more recent information will be available on the World Wide Web. In particular, reports and publications from governmental and statutory bodies should be consulted.

Albert, R. C. (1988). The historical context of water quality management for the Delaware estuary. *Estuaries*, **11**, 99–107.

Australian and New Zealand Environment and Conservation Council, and Agriculture and Resource Management Council of Australia and New Zealand (2000). *An Introduction to the Australian and New Zealand Guidelines for Fresh and Marine Water Quality*. Australian Water Association, Artarmon, NSW, Australia, 24 pp.

Bingham, N., Blowers, A. and Belshaw, C. (eds.) (2003). *Contested Environments*. John Wiley and Sons, Chichester and Open University, Milton Keynes, 344 pp.

Ducrotoy, J.-P. and Elliott, M. (1997). Interrelations between science and policy-making: the North Sea example. *Marine Pollution Bulletin*, **34**, 686–701.

Ducrotoy, J.-P., Elliott, M. and De Jonge, V. N. (2000). The North Sea. *Marine Pollution Bulletin*, **41**, 5–23.

Elliott, M. and Lawrence, A. J. (1998). The protection of species versus habitats – dilemmas for marine scientists. *Marine Pollution Bulletin*, **36**, 174–6.

Elliott, M., Fernandes, T. F. and De Jonge, V. N. (1999). The impact of recent European Directives on estuarine and coastal science and management. *Aquatic Ecology*, **33** 311–21.

Elliott, M. and De Jonge, V. N. (2002). The management of nutrients and potential eutrophication in estuaries and other restricted water bodies. *Hydrobiologia*, **475/476**, 513–24.

Fernandes, T. F., Elliott, M. and Da Silva, M. C. (1995). The management of European estuaries: a comparison of the features, controls and management framework of the Tagus (Portugal) and Humber (England). *Netherlands Journal of Aquatic Ecology*, **29**, 459–68.

Gubbay, S. (ed.) (1995). *Marine Protected Areas*. Chapman and Hall, London, 232 pp.

Hinchcliffe, S., Blowers, A. and Freeland, J. (eds.) (2003). *Understanding Environmental Issues*. John Wiley and Sons, Chichester and Open University, Milton Keynes, 179 pp.

King, M. (1995). *Fisheries Biology, Assessment and Management*. Fishing News Books, Oxford, 341 pp.

McEldowney, J. F. and McEldowney, S. (1996). *Environment and the Law: An Introduction for Environmental Scientists and Lawyers*. Longman, Harlow, Essex, 327 pp.

MAFF (Ministry of Agriculture, Fisheries and Food) (1993). *Coastal Defence and the Environment: A Strategic Guide*. Ministry of Agriculture, Fisheries and Food (now DEFRA), London.

Morris, R., Freeland, J., Hinchcliffe, S. and Smith, S. (eds.) (2003). *Changing Environments*. John Wiley and Sons, Chichester and Open University, Milton Keynes.

National Research Council (1994). *Restoring and Protecting Marine Habitats: the role of engineering and technology*. National Academy press, Washington DC, 193 pp.

OSPAR (Oslo and Paris Commission) (2000). *Quality Status Report 2000*. OSPAR Commission, London (and component subregional reports), 108 pp.

Penning-Rowsell, E. *et al.* (1992). *The Economics of Coastal Management*. Belhaven Press, London, 380 pp.

Ray, G. C. and McCormick-Ray, J. (2004). *Coastal-Marine Conservation: Science and policy*. Blackwell Publishing, Malden, MA, 327 pp.

Read, S. J., Elliott, M. and Fernandes, T. F. (2001). The possible implications of the Water Framework Directive and the Species and Habitats Directive on the Management of Marine Aquaculture. In P. A. Read *et al.* (eds.), *The Implications of Directives, Conventions and Codes of Practice on the Monitoring and Regulation of Marine Aquaculture in Europe (MARAQUA)*. EU FAIR Programme, PL98-4300 (1999–2000), Published by Fisheries Research Services, Aberdeen, p. 58–74.

Rees, J. A. (1990). *Natural Resources – Allocation, Economics and Policy*, 2nd Edition. Routledge, London, 499 pp.

Sayers, D. R. (1986). Derivation and application of environmental quality objectives and standards to discharges to the Humber estuary (U.K.). *Water Science and Technology*, **18**, 277–85.

Sunkin, M., Ong, D. M. and Wight, R. (2002). *Sourcebook on Environmental Law*. 2nd Edition. Cavendish Publ. Ltd, London, 901 pp.

Tietenberg, T. (2001). *Environmental Economics and Policy*. 3rd Edition, Harper Collins, New York, 498 pp.

Turner, R. K., Bateman, I. J. and Adger, W. M. (eds.) (2001). Economics of coastal and water resources: valuing environmental functions. Kluwer, Dordrecht, 342 pp.

Zedler, J. B. (ed.) (2001). *Handbook for Restoring Tidal Wetlands*. CRC Press, Boca Raton, Florida, 439 pp.

Index